# POLYMER PRODUCTS AND CHEMICAL PROCESSES

## Techniques, Analysis and Applications

# POLYMER PRODUCTS AND CHEMICAL PROCESSES

## Techniques, Analysis and Applications

*Edited by*
**Richard A. Pethrick, PhD, DSc, Eli M. Pearce, PhD,
and Gennady E. Zaikov, DSc**

Apple Academic Press

TORONTO    NEW JERSEY

Apple Academic Press Inc. | Apple Academic Press Inc.
3333 Mistwell Crescent | 9 Spinnaker Way
Oakville, ON L6L 0A2 | Waretown, NJ 08758
Canada | USA

©2014 by Apple Academic Press, Inc.

First issued in paperback 2021

*Exclusive worldwide distribution by CRC Press, a member of Taylor & Francis Group*

No claim to original U.S. Government works

ISBN 13: 978-1-77463-282-6 (pbk)
ISBN 13: 978-1-926895-53-6 (hbk)

**Library of Congress Control Number: 2013946170**

---

**Library and Archives Canada Cataloguing in Publication**

---

Polymer products and chemical processes: techniques, analysis and applications/edited by Richard A. Pethrick, PhD, DSc, Eli M. Pearce, PhD, and Gennady E. Zaikov, DSc.

Includes bibliographical references and index.
ISBN 978-1-926895-53-6
1. Polymers. 2. Polymerization. I. Pearce, Eli M., author, editor of compilation
II. Pethrick, R. A. (Richard Arthur), 1942-, author, editor of compilation
III. Zaikov, G. E. (Gennadiï Efremovich), 1935-, author, editor of compilation

TA455.P58P64 2013          668.9          C2013-905185-6

---

# ABOUT THE EDITORS

**Richard A. Pethrick, PhD, DSc**

Professor R. A. Pethrick, PhD, DSc, is currently a research professor and professor emeritus in the Department of Pure and Applied Chemistry at the University of Strathclyde, Glasglow, Scotland. He was Burmah Professor in Physical Chemistry and has been a member of the staff there since 1969. He has published over 400 papers and edited and written several books. Recently he has edited several publications concerned with the techniques for the characterization of the molar mass of polymers and also the study of their morphology. He currently holds a number of EPSRC grants and is involved with Knowledge Transfer Programmes involving three local companies involved in production of articles made out of polymeric materials. His current research involves AWE, and he has acted as a consultant for BAE Systems in the area of explosives as well as a company involved in the production of anticorrosive coatings.

Dr. Pethrick is on the editorial boards of several polymer and adhesion journals and was on the Royal Society of Chemistry Education Board. He is a Fellow of the Royal Society of Edinburgh, the Royal Society of Chemistry, and the Institute of Materials, Metals and Mining. Previously he chaired the 'Review of Science Provision 16-19' in Scotland and the restructuring of the HND provision in Chemistry. He was involved in the creation of the revised regulations for accreditation by the Royal Society of Chemistry of the MSci level qualifications in Chemistry. He was for a number of years the Deputy Chair of the EPSRC IGDS panel and involved in a number of reviews of the courses developed and offered under this program. He has been a member of the review panel for polymer science in Denmark and Sweden and the National Science Foundation in the USA.

**Eli M. Pearce, PhD**

Dr. Eli M. Pearce was President of the American Chemical Society. He served as Dean of the Faculty of Science and Art at Brooklyn Polytechnic

University in New York, as well as a professor of chemistry and chemical engineering. He was Director of the Polymer Research Institute, also in Brooklyn. At present, he consults for the Polymer Research Institute. A prolific author and researcher, he edited the *Journal of Polymer Science* (Chemistry Edition) for 25 years and was an active member of many professional organizations.

**Gennady E. Zaikov, DSc**

Gennady E. Zaikov, DSc, is Head of the Polymer Division at the N. M. Emanuel Institute of Biochemical Physics, Russian Academy of Sciences, Moscow, Russia, and Professor at Moscow State Academy of Fine Chemical Technology, Russia, as well as Professor at Kazan National Research Technological University, Kazan, Russia.

He is also a prolific author, researcher, and lecturer. He has received several awards for his work, including the the Russian Federation Scholarship for Outstanding Scientists. He has been a member of many professional organizations and on the editorial boards of many international science journals.

# CONTENTS

# LIST OF CONTRIBUTORS

**A. Akhalkatsi**
Institute of Physics of Tbilisi State University, Tamarashvil Str.6, Tbilisi 0186, Georgia

**V. Z. Aloev**
Kabardino-Balkarian State Agricultural Academy, Nal'chik – 360030, Tarchokov st., 1 a, Russian Federation

**M. P. Anachkov**
Institute of Catalysis, Bulgarian Academy of Sciences, Acad. G. Bonchev bl.11, 1113 Sofia, Bulgaria, E-mail: rakovsky@ic.bas.bg

**J. Aneli**
Institute of Machine Mechanics, 10 Mindeli Str., Tbilisi 0186, Georgia

**Sh. Arbab**
University of Guilan, P.O. Box 41635-3756, Rasht, Iran

**M. B. Belonenko**
Volgograd Institute of Business, Laboratory of Nanotechnologies, Volgograd, Russia

**Yu. V. Berestneva**
Donetsk National University, 24 Universitetskaya Street, 83 055 Donetsk, Ukraine
Christian Blobner
Fraunhofer Institute for Factory Operation and Automation IFF, Sandtorstrasse 22, 39106 Magdeburg, Germany

**V. V. Chernova**
The Bashkir State University, 32 Zaki Validy Str., Ufa, the Republic of Bashkortostan, 450074, Russia
Ina Ehrhardt
Fraunhofer Institute for Factory Operation and Automation IFF, Sandtorstrasse 22, 39106 Magdeburg, Germany

**A. K. Haghi**
University of Guilan, P.O. Box 41635-3756, Rasht, Iran

**M. Hasanzadeh**
Department of Textile Engineering, University of Guilan, Rasht, Iran, Tel.: +98-21-33516875; Fax: +98-182-3228375; E-mail: m_hasanzadeh@aut.ac.ir

**E. Chris Kazala**
Department of Agricultural, Food & Nutritional Science, University of Alberta, 4-10 Agriculture/Forestry Centre, Edmonton, Alberta, Canada, T6G 2P5

**G. Kirshenbaum**
Brooklyn Polytechnic University, 333 Jay str., Sixth Metrotech Center, Brooklyn, New York, NY, USA, E-mail: GeraldKirshenbaum@yahoo.com

**S. V. Kolesov**
The Institute of Organic Chemistry of the Ufa Scientific Center of the Russian Academy of Science, 71 October Prospect, Ufa, the Republic of Bashkortostan, 450054, Russia

**Paul P. Kolodziejczyk**
Biolink Consultancy Inc., P.O. Box 430, New Denver, B.C., Canada, V0G 1S0

**G. V. Kozlov**
Institute of Applied Mechanics of Russian Academy of Sciences, Leninskii pr., 32 a, Moscow 119991, Russian Federation, E-mail: IAM@ipsun.ras.ru

**E. I. Kulish**
The Bashkir State University, 32 Zaki Validy Str., Ufa, the Republic of Bashkortostan, 450074, Russia, E-mail: alenakulish@rambler.ru

**N. G. Lebedev**
Volgograd State University, Volgograd, Russia

**A. M. Lipanov**
Institute of Mechanics, Ural Branch of the Russian Academy of Sciences, T. Baramsinoy 34, Izhevsk, Russia, E-mail: postmaster@ntm.udm.ru

**G. M. Magomedov**
Dagestan State Pedagogical University, Makhachkala – 367003, Yaragskii st., 57, Russian Federation

**G. Mamniashvili**
Institute of Physics of Tbilisi State University, Tamarashvil Str.6, Tbilisi 0186, Georgia

**B. Hadavi Moghadam**
Department of Textile Engineering, University of Guilan, Rasht, Iran

**Vahid Mottaghitalab**
Textile Engineering Department, Faculty of Engineering, University of Guilan, Rasht P.O. BOX 3756, Guilan, Iran

**L. Nadareishvili**
Institute of Cybernetics of Georgian Technical University, 5 S.Euli Str., Tbilisi 0186,Georgia

**B. Noroozi**
University of Guilan, P.O. Box 41635-3756, Rasht, Iran

**I. A. Opeida**
L.M. Litvinenko Institute of Physical Organic and Coal Chemistry National Academy of Sciences of Ukraine. 70, R.Luxemburg Str., 83 114, Donetsk, Ukraine

**E. N. Pasternak**
Donetsk National University, 24 Universitetskaya Street, 83 055 Donetsk, Ukraine

**E. M. Pearce**
Brooklyn Polytechnic University, 333 Jay str., Sixth Metrotech Center, Brooklyn, New York, NY, USA, E-mail: EPearce@poly.edu

**Xiao Qiu**
Department of Food and Bioproduct Sciences, University of Saskatchewan, 51 Campus Drive, Saskatoon, Saskatchewan, Canada, S7N 5A8

**S. Rafiei**
University of Guilan, P.O. Box 41635-3756, Rasht, Iran

**S. K. Rakovsky**
Institute of Catalysis, Bulgarian Academy of Sciences, Acad. G. Bonchev bl.11, 1113 Sofia, Bulgaria, E-mail: rakovsky@ic.bas.bg

**E. V. Raksha**
Donetsk National University, 24 Universitetskaya Street, 83 055 Donetsk, Ukraine

**E. H. Michael Schenk**
Fraunhofer Institute for Factory Operation and Automation IFF, Sandtorstrasse 22, 39106 Magdeburg, Germany

**Holger Seidel**
Fraunhofer Institute for Factory Operation and Automation IFF, Sandtorstrasse 22, 39106 Magdeburg, Germany

**Saleh Shah**
Alberta Innovates-Technology Futures, P.O. Bag 4000, Vegreville, Alberta, Canada, T9C 1T4

**Anamika Singh**
Division of Reproductive and Child Health, Indian Council of Medical Research, New Delhi

**Rajeev Singh**
Division of Reproductive and Child Health, Indian Council of Medical Research, New Delhi, E-mail: 10rsingh@gmail.com

**Crystal L. Snyder**
Department of Agricultural, Food & Nutritional Science, University of Alberta, 4-10 Agriculture/Forestry Centre, Edmonton, Alberta, Canada, T6G 2P5

**S. A. Sudorgin**
Volgograd State University, Volgograd, Russia, E-mail: sergsud@mail.ru

**N. A. Turovskij**
Donetsk National University, 24 Universitetskaya Street, 83 055 Donetsk, Ukraine; E-mail: N.Turovskij@donnu.edu.ua

**A. V. Vakhrushev**
Institute of Mechanics, Ural Branch of the Russian Academy of Sciences, T. Baramsinoy 34, Izhevsk, Russia, E-mail: postmaster@ntm.udm.ru

**V. P. Volodina**
The Institute of Organic Chemistry of the Ufa Scientific Center of the Russian Academy of Science, 71 October Prospect, Ufa, the Republic of Bashkortostan, 450054, Russia

**Randall J. Weselake**
Department of Agricultural, Food & Nutritional Science, University of Alberta, 4-10 Agriculture/Forestry Centre, Edmonton, Alberta, Canada, T6G 2P5

**Kh. Sh. Yakh'yaeva**
Dagestan State Pedagogical University, Makhachkala – 367003, Yaragskii st., 57, Russian Federation

**Yu. G. Yanovskii**
Institute of Applied Mechanics of Russian Academy of Sciences, Leninskii pr., 32 a, Moscow 119991, Russian Federation, E-mail: IAM@ipsun.ras.ru

**G. E. Zaikov**
N.M. Emanuel Institute of Biochemical Physics, Russian Academy of Sciences, 4 Kosygin str., Moscow 119334 Russia, E-mail: chembio@sky.chph.ras.ru

**Z. M. Zhirikova**
Kabardino-Balkarian State Agricultural Academy, Nal'chik – 360030, Tarchokov st., 1 a, Russian Federation

**M. Yu. Zubritskij**
L.M. Litvinenko Institute of Physical Organic and Coal Chemistry National Academy of Sciences of Ukraine, 70, R.Luxemburg Str., 83 114, Donetsk, Ukraine

# LIST OF ABBREVIATIONS

| | |
|---|---|
| AC | activated complex |
| AdometDC | S. adenosylmenthionine decarboxylase |
| AI | aromatization index |
| ALA | α-linolenic acid |
| BBIC | bioluminescent bioreporter integrated circuit |
| BSR | butadiene-stirene rubber |
| CANARY | Cellular Analysis and Notification of Antigen Risks and Yields |
| CdTe | cadmium telluride |
| CIGS | Copper Indium/Gallium Di Selenide |
| CNT | carbon nanotubes |
| CVD | chemical deposition method |
| DCI | disubstituted CI |
| DCP | dicumeneperoxide |
| DHA | docosahexaenoic acid |
| DMF | dimethylformamide |
| DSC | differential scanning calorimetery |
| DSSC | dye-sensitized solar cells |
| EDG | electron-donating groups |
| EFA | essential fatty acids |
| EPA | eicosapentaenoic acid |
| EWG | electron-withdrawing groups |
| FF | fill factor |
| FTIR | Fourier transform infrared spectroscopy |
| GFP | green fluorescent protein |
| GPi | graphitized polyimide |
| HAART | highly active retroviral therapy |
| IRMOFs | isoreticular metal-organic frameworks |
| ITO | indium tin oxide |
| LA | linoleic acid |

| | |
|---|---|
| MCI | monosubstituted CI |
| MMD | mass-molecular distribution |
| MOFs | metal-organic frameworks |
| MPP | maximum power point |
| MWCNTs | multi walled carbon nanotubes |
| ONRL | Oak Ridge National Laboratory |
| PAHs | polycylic aromatic hydrocarbons |
| PAN | polyacrylonitrile |
| PAni | polyaniline |
| PET | poly thyleneterphthalate |
| PKS | polyketide synthase |
| PO | primary ozonide |
| PP | poly propylene |
| PSM | post-synthetic modification |
| PV | photovoltaic |
| PZT | Plumbum Zirconate Titanate |
| RF | remote frequency |
| RT | reverse transcriptase |
| SBUs | secondary building units |
| SDA | stearidonic acid |
| SEM | scanning electron microscopy |
| TC | technical carbon |
| TSC | textile solar cell |
| UV | ultraviolet |
| VLCPUFA | very long chain polyunsaturated fatty acids |
| XRD | X-ray diffraction |

# LIST OF SYMBOLS

| | |
|---|---|
| $a$ | lower linear scale of fractal behavior |
| $c$ | concentration of the monomeric units |
| $d$ | Euclidean space |
| $D_f$ | fractal dimension |
| $E_a$ | activation energy |
| $E_{max}$ | maximum field amplitude |
| $E_n$ and $E_m$ | elasticity moduli of nanocomposite and matrix polymer |
| F | free energy |
| $I_{sc}$ | short-circuit current |
| $k$ | Boltzmann constant |
| $l_0$ | skeletal bond length |
| $l_n$ | linear length scale |
| $N$ | number of particles with size $\rho$ |
| $P_{in}$ | incident light power density |
| $P_{max}$ | maximum electrical power |
| $R_g$ | gyration radius |
| $R_p$ | nanofiller particle |
| $S_i$ | quadrate area |
| $S_u$ | nanoshungite particles specific surface |
| $t$ | ozonation time (s) |
| $T_g$ | glass transition temperature |
| $V_{oc}$ | open-circuit voltage |
| Y | total ozonide yield |

## Greek Symbols

| | |
|---|---|
| $\alpha$ | Henry's coefficient |
| $\gamma$ | globe swelling |
| $\gamma_L$ | Grüneizen parameter |
| $\eta_{rel}$ | relative viscosity |

| | |
|---|---|
| $\varphi$ | reacted ozone |
| $\varphi_{if}$ | interfacial regions |
| $\nu$ | Poisson ratio |
| $\rho_n$ | nanofiller particles aggregate density |

# PREFACE

*Polymer Products and Chemical Processes* presents leading-edge research in the rapidly changing and evolving field of polymer science as well as on chemical processing. The topics in the book reflect the diversity of research advances in the production and application of modern polymeric materials and related areas, focusing on the preparation, characterization, and applications of polymers. The book also covers various manufacturing techniques. The book will help to fill the gap between theory and practice in industry.

*Polymer Products and Chemical Processes* is a collection of 11 chapters that highlights many important areas of current interest in polymer products and chemical processes. It is also gives an up-to-date and thorough exposition of the present state of the art of polymer analysis. There are many chapters that will familiarizes the reader with new aspects of the techniques used in the examination of polymers, including chemical, physico-chemical and purely physical methods of examination.

In chapter 1, some new aspects of ozone and its reactions on diene rubbers are presented. The importance of nanocomposites in today's modern science is highlighted in chapter 2, in which different types of polymer nanocomposites structures are studied in detail. The simulation of nanoelements' formation and interaction is explained in chapter 3. Chapter 4 is divided into three sections to introduce new points of views on advanced polymers. The stabilization process of PAN nanofibers is studied in detail in chapter 5. In chapter 6, carbon nanotubes' structure in polymer nanocomposites is updated for our readers. Exploring the potential of oilseeds as a sustainable source of oil and protein for aquaculture feed is presented in chapter 7. Microbial biosensors are introduced in chapter 8. New development of solar cloth by electrospinning technique is well defined in chapter 9. Applications of metal-organic frameworks in textiles are described in chapter 10 and chapter 11 and are divided into 3 sections in present important topics related to the book's objectives.

This book describes the types of techniques now available to the polymer chemist and technician, and discusses their capabilities, limitations, and applications and provides a balance between materials science and mechanics aspects, basic and applied research, and high technology and high volume (low cost) composite development

— **Richard A. Pethrick, PhD, DSc, Eli M. Pearce, PhD,**
**and Gennady E. Zaikov, DSc**

# CHAPTER 1

# SOME NEW ASPECTS OF OZONE AND ITS REACTIONS WITH DIENE RUBBERS

G. E. ZAIKOV, S. K. RAKOVSKY, M. P. ANACHKOV, E. M. PEARCE, and G. KIRSHENBAUM

## CONTENTS

## 1.1   INTRODUCTION

The interest in the reaction of ozone with polydienes is due mainly to the problems of ozone degradation of rubber materials [1–4] and the application of this reaction to the elucidation of the structures of elastomers [5–8]. It is also associated with the possibilities of preparing bifunctional oligomers by partial ozonolysis of some unsaturated polymers [9–12]. Usually the interpretation of experimental results are based on a simplified scheme of Criegee's mechanism of C=C-double bond ozonolysis, explaining only the formation of the basic product – ozonides [13, 14].

The reactions of ozone with 1,4-*cis*-polybutadiene (SKD); Diene 35 NFA (having the following linking of the butadiene units in the rubber macromolecules: 1,4-*cis* (47%), 1,4-*trans* (42%), 1,2-(11%); 1,4-*cis*-polyisoprene (Carom IR 2200), 1,4-trans-polychloroprene (Denka M 40), and 1,4-*trans*-polyisoprene have been investigated in $CCl_4$ solutions. The changes of the viscosity of the polymer solutions during the ozonolysis have been characterized by the number of chain scissions per molecule of reacted ozone ($\varphi$). The influence of the conditions of mass-transfer of the reagents in a bubble reactor on the respective $\varphi$ values has been discussed. The basic functional groups-products from the rubbers ozonolysis have been identified and quantitatively characterized by means of IR-spectroscopy and $^1$H-NMR spectroscopy. A reaction mechanism, that explains the formation of all identified functional groups, has been proposed. It has been shown that the basic route of the reaction of ozone with elastomer double bonds – the formation of normal ozonides does not lead directly to a decrease in the molecular mass of the elastomer macromolecules, because the respective 1,2,4-trioxolanes are relatively stable at ambient temperature. The most favorable conditions for ozone degradation emerge when the cage interaction between Criegee intermediates and respective carbonyl groups does not proceed. The amounts of measured different carbonyl groups have been used as an alternative way for evaluation of the intensity and efficiency of the ozone degradation. The thermal decomposition of partially ozonized diene rubbers has been investigated by DSC. The respective values of the enthalpy, the activation energy and the reaction order of the 1, 2, 4-trioxolanes have been determined.

In most cases, quantitative data on the functional groups formed during the reaction are missing [15–18]. At the same time alternative conversion routes of Criegee's intermediates, which lead to the formation of carbonyl compounds and some other so called "anomalous products" of the ozonolysis, are of great importance for clarifying the overall reaction mechanism [19–21]. The mechanism of ozone degradation of rubbers is also connected with the nonozonide routes of the reaction, because the formation of the basic product of ozonolysis, normal ozonide, does not cause any chain scission and/or macromolecule cross-linking [22].

In this work the changes in the molecular mass of different types of diene rubbers during their partial ozonolysis in solution have been investigated. By means of IR and $^1$H-NMR spectroscopy ozonolysis products of the elastomers have been studied. The effects of the nature of the double bond substituents and its configuration on the degradation mechanism have been considered. By using differential scanning calorimetry the thermal decomposition of the functional groups of peroxide type has also been investigated.

## 1.2  EXPERIMENTAL METHODS

### 1.2.1  MATERIALS

Commercial samples of 1,4-*cis*-polybutadiene (SKD; E-BR); polybutadiene (Diene 35 NFA; BR); 1,4-*cis*-polyisoprene (Carom IR 2200; E-IR) and polychloroprene (Denka M 40; PCh) were used in the experiments (Table 1). The 1, 4-*trans*-polyisoprene samples were supplied by Prof. A. A. Popov, Institute of Chemical Physics, Russian Academy of Sciences. All rubbers were purified by threefold precipitation from CCl$_4$ solutions in excess of methanol. The above mentioned elastomer structures were confirmed by means of $^1$H-NMR spectroscopy. *Ozone* was prepared by passing oxygen flow through a 4–9 kV electric discharge.

**TABLE 1**   Some characteristics of polydiene samples.

| Elastomer | Monomeric unit | Unsaturation degree,% | 1,4-*cis*, % | 1,4-*trans*-, % | 1,2-, % | 3,4-, % | $M_v.10^{-3}$ | n |
|---|---|---|---|---|---|---|---|---|
| SKD | -CH=CH- | 95–98 | 87–93 | 3–8 | 3–5 | – | 454 | 2.1 |
| Diene 35 NFA | -CH=CH- | 97 | 47 | 42 | 11 | | 298 | 2.63 |
| Carom IR 2200 | -C(CH₃)=CH- | 94–98 | 94–97 | 2–4 | – | 1–2 | 380 | 2.0 |
| 1,4-*trans* PI | -C(CH₃)=CH- | 95–97 | | 95–97 | | | 310 | 2.3 |
| Denka M40 | -C(Cl)=C- | 94–98 | 5 | 94 | – | – | 180 | 1.8 |

## 1.2.2   OZONATION OF THE ELASTOMER SOLUTIONS

The ozonolysis of elastomers was performed by passing an ozone-oxygen gaseous mixture at a flow rate of $v=1.6\times10^{-3}\pm0.1$ l.s⁻¹ through a bubbling reactor, containing 10–15 ml of polymer solution (0.5–1 g in $CCl_4$) at 293 K. Ozone concentrations in the gas phase at the reactor inlet ($[O_3]_i$) and outlet ($[O_3]_u$) were measured spectrophotometrically at 254 nm [23]. The amount of consumed ozone (G, mole) was calculated by the Eq. (1):

$$G=v([O_3]_i.[O_3]_u)t \qquad (1),$$

Where $t$ is the ozonation time (s). The degree of conversion of the double C=C bonds was determined on the basis of the amount of reacted ozone and the reaction stoichiometry [23].

   *Note*: $m_v$ is the average molecular weight, determined viscosimetrically from equation $[\eta]=k.M_v^{\alpha'}$, where $[\eta]=(\eta_1/C)(1+0.333\eta_1)$, $\eta_1=\eta_{rel-}1$, $\eta_{rel-}$ is the intrinsic viscosity; C-solution concentration; $k=1.4.10^{-4}$ – Staudinger's constant and $\alpha' = 0.5$–1.5 – constant depending on the rubber type, being one for natural rubber; $M_v\approx M_w$; $n=M_w/M_n$, where $M_w$ and $M_n$ are the average weight and number average molecular mass, respectively [22].

## 1.3   RESULTS AND DISCUSSION

Florry [24] has shown that the reactivity of the functional groups in the polymer molecule does not depend on its length. It is also known that some reactions of the polymers proceed more slowly, compared with their low molecular analogs (catalytic hydrogenation). The folded or unfolded forms of the macromolecules provide various conditions for contact of the reagents with the reacting parts [4, 25]. By using the modified version of this principle [26] it was possible to explain the proceeding of reactions without any specific interactions between the adjacent C=C bonds and the absence of diffusion limitations. The study of the mass-molecular distribution (MMD) is in fact a very sensitive method for establishing the correlation between molecular weight ($M._w.$) and the reactivity. The theory predicts that the properties of the system: polymer-solvent can be described by the parameter of so called globe swelling ($\gamma$), which defines the free energy (F) of the system and thus the rate constant of the reaction. For a reversible reaction, i.e., polymerization – depolymerization, the dependence of the rate constant of the chain length growing on the molecular weight is expressed by the following equation:

$$\ln k_{pj}/k_p = - \text{const.}(5\gamma - 3/\gamma).(d\gamma/dM).M_0 \qquad (2),$$

Where $m_0$ is the molecular weight of the studied sample and $k_p$ is the rate constant for infinitely long macromolecules. A good correlation between the theoretical and experimental data for polystirene solutions in benzene has been found in Ref. [27].

The study of the polymer degradation is complicated by their structural peculiarities on molecular and supramolecular level and diffusion effects. It is difficult to find simple model reactions for clarification of particular properties and for the express examination of the proposed assumptions. An exception in this respect is the ozone reaction with C=C bonds, whose mechanism has been intensively studied and could be successfully applied upon ozonolysis of polymeric materials [28].

Table 2 summarises the rate constants of the ozone reactions with some conventional elastomers and polymers and their low molecular analogs, synthesized by us. It is seen that the reactivities of elastomers and polymers and their corresponding low molecular analogs, as it is demonstrated by their rate constants, are quite similar, thus suggesting similar mechanisms of their reaction with ozone. This statement is also confirmed by: (1) the dependence of $k$ on the inductive properties of substituents: for example $k$ of polychloroprene is higher than that of vinylchloride due to the presence of two donor substituents; and (2) the dependence of $k$ on the configuration of the C=C bond in *trans*-isomer (gutta-percha) and *cis*-isomer (natural rubber).

**TABLE 2**   Rate constants of ozone reactions with polymers and low molecular analogs in $CCl_4$, 20°C.

| Compound | M.w. | $k.10^{-4}, M^{-1}.s^{-1}$ |
|---|---|---|
| Polychloroprene | $8.10^5$ | $0.42\pm0.1$ |
| Vinylchloride | 62.45 | 0.18 |
| 2-bromopropene | 121 | $0.28\pm0.05$ |
| Polybutadiene | $3.3.10^5$ | $6.0\pm1$ |
| Cyclododecatriene-1,5,9 | 162 | $35\pm10$ |
| Poly(butadiene-costirene) | $8.10^4$ | $6\pm1$ |
| Gutta-percha | $3.10^4$ | $27\pm5$ |
| Natural rubber | $1.10^6$ | $44\pm10$ |
| 2-me-pentene-2 | 85 | $35\pm10$ |
| Squalene | 410 | $74\pm15$ |
| Polystirene | $5.10^5$ | $0.3.10^{-4}$ |
| Cumene | 120 | $0.6.10^{-4}$ |
| Polyisobutylene | $1.7.10^5$ | $0.02.10^{-4}$ |
| Cyclohexane | 84 | $0.01.10^{-4}$ |

    It has been found out that the effects, related either to the change in the macromolecule length or to the folding degree, do not affect the ozonolysis in solution. Probably this is due to the fact that the reaction is carried out in elastomeric solutions, in which the macromolecules are able to do free intramolecular movements and they do not react with adjacent macromolecules. Moreover, the rate of macromolecules reorganization is probably higher than the rate of their reaction with ozone as the experiment does not provide any evidence for the effects of the change in the parameters pointed earlier [29].

However, it should be noted that $k$ values of the elastomers are about 2–6 times lower that those of the low molecular analogs. The accuracy of activation energy $(E_a)$ determination does not allow to estimate the contribution of the two parameters: preexponential factor (A) or $E_a$ for the decrease in $k$. If we assume that the mechanism of ozone reaction with monomers and elasomers is similar, i.e., the reactions are isokinetic, then $A_{mon} = A_{pol}$. At $k_{mon}/k_{pol} = 2 \div 6$ the difference in $E_a$ at 20°C will be 0.5–1.0 kcal/mole. At the low experimental values of $E_a$, these differences will become commensurable and thus the determination of $E_a$ is not sufficiently accurate. In this case two assumptions could be made which can give a reasonable explanation for the lower values of $k_{pol}$: (1) the reorientation of the macromolecules is a slower process that that of olefins, which would results in $A_{pol}$ lower than $A_{mon}$; and (2) the addition of ozone to C=C bonds is accompanied by the rehybridization of the C-atoms from $sp^2-sp^3$ and the movements of the polymer susbstituents during the formation of activated complex (AC) will be more restricted than those in olefins, mainly because of their greater molecular mass and sizes. This will ultimately result in decrease of the rate constant.

Table 2 shows some examples of ozonolysis of saturated polymers – polystyrene and polyisobutylene. These reactions take place not via the mechanism of ozone reaction with the double bonds but through a hidden radical mechanism with rate constants of 4–5 orders of magnitude lower.

### 1.3.1   POLYBUTADIENES

Because of the high viscosity and high value of rate constants the reaction takes place either in the diffusion or in the mixed region. In order to obtain correct kinetic data we have used the theory of boundary surface [30]:

$$[O_3] = \alpha[O_3]_0 . \exp[-\delta(k.c.D)^{1/2}], \tag{3}$$

where $[O_3]$ is the ozone concentration at a distance $\delta$; $\alpha$ – Henry's coefficient; $[O_3]_0$ – equilibrium ozone concentration in the gas phase at the reactor inlet; $\delta$ – penetration depth of ozone from the interphase surface [22]; $k$ – rate constant of the ozone reaction with double bonds; $c$ – concentration of the monomeric units; $D$ – diffusion coefficient of ozone in the liquid phase.

It was found out that the relative viscosity decreases exponentially upon ozonation of SKD solutions (Fig. 1). As the viscosity is proportional to the molecular weight it follows that the polydiene consumption should be described by first or pseudo first order kinetics.

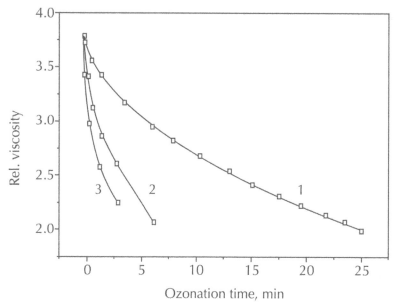

**FIGURE 1** Dependence of the relative viscosity ($\eta_{rel}$) of SKD solutions (0.6 g in 100 ml $CCl_4$) on reaction time at ozone concentrations of: $1-1.10^{-5}$ M; $2-4.5.10^{-5}$ M; $3-8.25.10^{-5}$ M.

The value of $\phi$, corresponding to the number of degraded polymeric molecules per one absorbed ozone molecule can be used to calculate the degradation efficiency. The value of this parameter ($\phi$) may be estimated using the following equation:

$$\phi = 0.5 \left[ (M_{vt})^{-1} - (M_{v0})^{-1} \right].P/G, \qquad (4)$$

where $m_{vt}$ is the molecular weight at time moment T; $M_{vo}$ – the initial molecular weight; P – the polymer amount; G – amount of consumed ozone.

The dependence of $\phi$ on G is a straight line for a given reactor and it depends on the hydrodynamic conditions in the reactor. It is seen from Fig. 2 that the $\phi$ values are increasing linearly with the reaction time and decreasing with increase in ozone concentration. The corresponding dependences for Carom IR 2200 and Denka M40 ozonolysis are similar. The $\phi$ values for G®0 were used to avoid the effect of hydrodynamic factors on them.

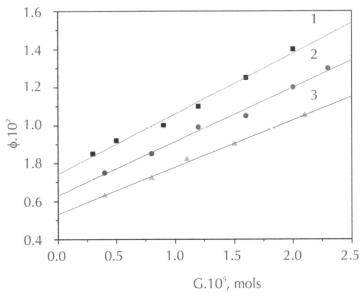

**FIGURE 2**  Dependence of $\phi$ on G for SKD (0.6/100) at various ozone concentrations: $1-1.10^{-5}$ M; $2-4.5.10^{-5}$ M; $3-8.25.10^{-5}$ M.

The values of $\phi$ found for SKD, Carom IR 2200 and Denka M40 at $[O_3] = 1.10^{-5}$ M amount to $0.7.10^{-2}$, $0.78.10^{-2}$ and $0.14$, respectively, and the slopes are: $-40$, $-70$ and $200$ $M^{-1}$, respectively. Substituting with the known values for the parameters in Eq. (3) we have obtained $\delta$ within the range of $1.10^{-3}-2.10^{-4}$ cm, which indicates that the reaction is taking place in the volume around the bubbles, and hence in the diffusion region.

The ozonolysis of polydienes in solutions is described by the Criegee's mechanism. The C=C bonds in the macromolecules are isolated as they are separated by three simple C-C bonds. According to the classical concepts, the C=C bonds configuration and the electronic properties of the groups bound to them, also affect the polymer reactivity; similarly they do this in case of the low molecular olefins. The only difference is that the polymer substituents at the C=C bonds are less mobile, which influences the $sp^2-sp^3$ transition and the ozonides formation. In the first stage, when primary ozonides (PO) (Scheme 1, reaction 1) are formed, the lower mobility of the polymer substituents requires higher transition energy, the rate being respectively lower, compared to that with low molecular olefins and the existing strain accelerates the PO decomposition to zwitterion and carbonyl compound. The lower mobility of the polymer parts impedes the further ozonide formation and causes the zwitterion to leave the cage and pass into the volume, which in its turn accelerates the degradation process. The latter is associated either with its monomolecular decomposition or with its interaction with low molecular components in the reaction mixture. The efficiency of degradation is determined by the C=C bonds location in the macromolecule, for example, at C=C bond location from the macromolecule center to its end, it is in the range from 2 to 1.

$$M_1 = (1/g).M_0, \tag{5}$$

where $m_2 = M_0-M_1$; $1 \leq g \leq 2$ – coefficient pointing the C=C-bond location; $M_0$, $M_1$ and $M_2$ – the molecular weights of the initial macromolecule and of the two degradated polymer parts, respectively.

At g=2, i.e., when the broken C=C bond is located in the macromolecule center, the values of $M_1$ and $M_2$ will be exactly equal to $M_0/2$, at g®1, i.e., at terminal C=C bond in the polymer chain, the value of $M_1$ will be approximated to $M_0$ and thus the value of $M_2$ will be practically insignificant. For example, $M_2$ may be 50–1000, which is 3–4 orders of magnitude less than that of the macromolecule and in fact degradation process will not occur. The viscosimetric determination of the molecular weight, which we have applied in our experiments, has accuracy of ±5% and does not allow the differentiation of molecular weights of 22,700, 19,000 and 9,000 for the corresponding types of rubbers. This suggests that the cleavage of C=C bonds, located at distances of 420, 280 and 100 units from the macromolecule end, would not affect the measured molecular weight.

Since the reaction of elastomers ozonolysis proceeds either in the diffusion or in diffusion-kinetic region, at low conversions each new gas bubble in the reactor would react with a new volume of the solution. On the other hand, the reaction volume is a sum of the liquid layers surrounding each bubble. It is known that the depth of the penetration from the gas phase into the liquid phase is not proportional to the gas concentration and thus the rise of ozone concentration would increase the reaction volume to a considerably smaller extent than the ozone concentration. This leads to the occurrence of the following process: intensive degradation processes take place in the microvolume around the bubble and one macromolecule can be degraded to many fragments, while the macromolecules out of this volume, which is much greater, may not be changed at all. Consequently with increase in ozone concentration, one may expect a reduction of coefficient MMD and increase in the oligomeric phase content. This will result in apparent decrease of f in case of the viscosimetric measurements. The discussion above enables the correct interpretation of the data in Fig. 3.

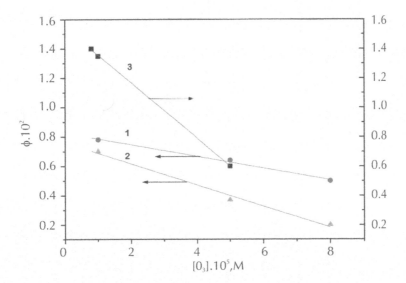

**FIGURE 3**   Dependence of f on ozone concentration for elastomer solutions: 1 – SKD (0.6/100); 2 – Carom IR 2200 (0.6/100); 3 – Denka M40 (1/100).

In the spectra of the ozonized polybutadienes the appearance of bands at $1111 \times 1735$ cm$^{-1}$, that are characteristic for ozonide and aldehyde groups, respectively, is observed [22, 31]. It was found out that the integral intensity of ozonide peak in the 1,4-*cis*-polybutadiene (E-BR) spectrum, is greater and that of the aldehyde is considerably smaller in comparison with the respective peaks in the Diene 35 NFA (BR) spectrum, at one and the same ozone conversion degree of the double bonds. The mentioned differences in the aldehyde yields indicate that, according to IR-analysis, the degradation efficiency of the BR solutions is greater.

The $^1$H-NMR spectroscopy provides much more opportunities for identification and quantitative determination of functional groups, formed during ozonolysis of polybutadienes [32]. Figure 4 shows spectra

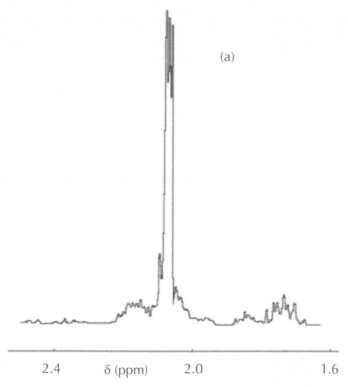

**FIGURE 4 (a–c)**    1H-250 MHz NMR spectra of E-BR solutions (0.89 g / 100 ml CCl4) ozonized to 18% conversion of the double bonds (external standard TMS; digital resolution 0.4 Hz, 20°C).

of ozonized E-BR. The signals of the ozonolysis products are decoded in Table 3 on the basis of Fig. 5. The ozonide: aldehyde ratio, determined from NMR spectra, was 89:11 and 73:27 for E-BR and BR, correspondingly. The peak at 2.81 ppm is present only in the spectra of ozonized Diene 35 NFA. It is usually associated with the occurrence of epoxide groups [33]. The integrated intensity of that signal compared to the signal of aldehyde protons at 9.70–9.79 ppm, was about 10%. Similar signal at 2.75 ppm has been registered in the spectra of ozonized butadiene-nitrile rubbers, where the 1,4-*trans* double bonds are dominant [31].

**TABLE 3**  Assignment of the signals in the $^1$H-NMR spectra of partially ozonized E-BR and BR rubbers.

| Assignment of the signals (according Fig. 4) | Chemical shifts (ppm) | | Literature |
|---|---|---|---|
| | **E-BR** | **BRA** | |
| a | 5.10–5.20 max 5.12, 5.16 | 5.05–5.18 max 5.10, 5.15 | [19, 34] |
| b | 1.67–1.79 max 1.72, 1.76 | 1.66–1.80 max 1.73 | [7, 19] |
| c | 9.75 | 9.74 | [33] |
| d | 2.42–2.54 max 2.47 | 2.42–2.54 max 2.50 | [33] |
| e | 2.27–2.42 max 2.35 | 2.27–2.42 max 2.35 | [33] |
| f | | max 2.81 | [33] |

According to [2, 10] two isomeric forms of 1, 2, 4-trioxolanes exist. The ratio between them is a function of the double bond stereochemistry, steric effect of the substituents and the conditions of ozonolysis. It was fond out only on the low molecular weight alkenes [19, 21]. The $^1$H-NMR spectroscopy is the most powerful method for determination of the *cis/trans* ratio of ozonides (in the case of polymers it is practically the only one method that can be applied). The measuring is based on the differences in the chemical shifts of the methine protons of the two isomers:

the respective signal of the *cis* form appears in lower field compared to the *trans* one [19, 21]. The multiplet on Fig. 3 in the area of 5.1–5.18 ppm could be interpreted as a result of partial overlapping of triplets of *trans* and *cis*-ozonides: 5.12 ppm (t, J»5 Hz, 2H) and 5.16 ppm (t, J»5 Hz, 2H), respectively. It is interesting to note that the *cis/trans* ratio of the E-BR 1, 2, 4-trioxolanes is practically equal to that obtained from *cis*-3-hexene [19, 34]. The resolution of the respective BR spectrum does not allow consideration in detail of the multiplicity of the signals at 5.10 and 5.15 ppm. In this case the area of the signals is widened, most probably due to the presence of ozonide signals of the 1, 2-monomer units [20].

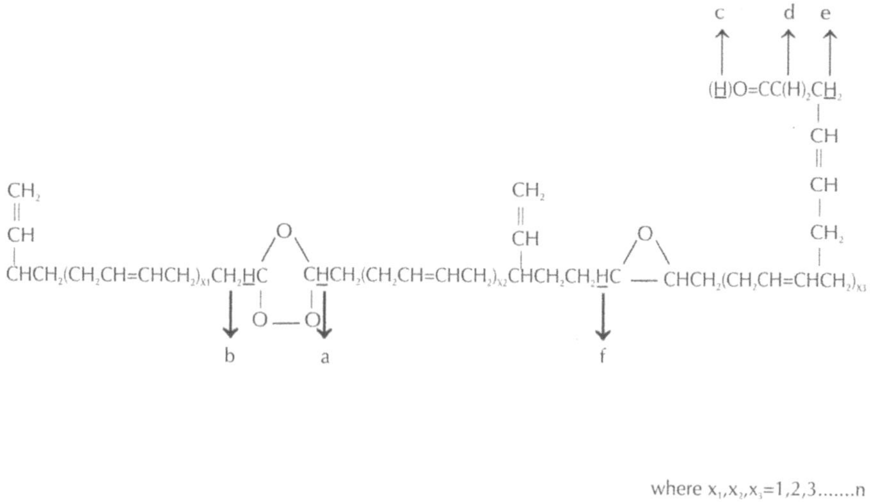

where $x_1, x_2, x_3 = 1, 2, 3 \ldots \ldots n$

**FIGURE 5** Selection of protons with characteristic signals in the 1H-250 MHz NMR spectra of partially ozonized polybutadiene macromolecules.

The basic route of the reaction – the formation of normal ozonides does not lead directly to a decrease in the molecular mass of the elastomer macromolecules, because the respective 1,2,4-trioxolanes are relatively stable at ambient temperature (by analogy with Scheme 1 of polyisoprenes, see it below) [31, 32]. The most favorable conditions for ozone degradation emerge when the cage interaction (Scheme 1, reaction 3) does not proceed. Therefore, the higher ozonide yield the lower the intensity of ozone degradation of the polybutadienes and vice versa. As it was already determined the ozonide yields for the 1,4-*cis*- and 1,2-monomer units are close to 83–90%, whereas that for the 1,4-*trans* units is about 50%. The amount of aldehyde groups is usually used for evaluation of the intensity and efficiency (number of chain scissions per molecule of reacted ozone) of ozone degradation of elastomers. In this case it should be taken into account that the dominant route of degradation leads to the formation of 1 mole of aldehyde from 1 mole of ozone [32].

### 1.3.2  POLYISOPRENES

The positive inductive effect of the methyl group in polyisoprene enhances the rate of ozone addition to the double bonds from $6.10^4$ for SKD to $4.10^5$ $M^{-1}.s^{-1}$ for Carom IR 2200. The infrared spectra of ozonized 1,4-*trans*-polyisoprene (Z-IR) show two intense bands at 1,100 and 1,725 cm⁻¹, characteristic of ozonide and keto groups, respectively [19, 35]. These spectra are identical with the well-known spectra of 1,4-*cis*-polyisoprenes (E-IR), as far as ozonide and carbonyl bands are concerned [22, 36]. It was found that the integrated intensity of the peak at 1100 cm⁻¹ in the E-IR and Z-IR spectra is equal for one and the same amount of reacted ozone. By analogy with the peak at 1,110 cm⁻¹, the integral intensity of the peak at 1,725 cm⁻¹ is also one and the same. The latter show that, according to the infrared spectra, the degradation efficiencies of E-IR and Z-IR with respect to the amount of consumed ozone do not practically differ.

The ¹H-NMR spectroscopy affords much more opportunities for identification and quantitative determination of functional groups formed on ozonolysis of polyisoprenes. Figure 6 shows spectra of nonozonized and ozonized E-IR. Changes in the spectra of ozonized elastomers are decoded in Table 4 on the basis of Fig. 7. It is seen that besides ketones, aldehydes are also formed as a result of ozonolysis. A comparison between methylene and methine proton signals of the nonozonized polyisoprenes and the corresponding ozonide signals indicates a considerable overlap in the ranges 1.60–1.90 and 4.5–5.5 ppm. A signal overlap is also registered in the 2.00–2.20 ppm region, characteristic of methyl protons of keto groups. Because of the reasons mentioned earlier, the ozonides and aldehydes were quantified by using the integrated intensity of the signals at 1.40 (a) and 9.70–9.79 (g) ppm, respectively. Ketone amounts were determined as a difference between the total intensity of the methylene signals from aldehydes and ketones, 2.40–2.60 and 2.35–2.60 ppm, respectively, and the doubled intensity of the aldehyde signal at 9.70–9.79 ppm. Thus the obtained ozonide: ketone: aldehyde ratio was 40:37:23 and 42:39:19 for 1,4-*cis*-polyisoprene and 1,4-*trans*-polyisoprene, correspondingly. The peak at 2.73 ppm, present in the spectra of both ozonized elastomers, is associated with the occurrence of epoxide groups [33, 37]. The integrated intensity of that signal compared to the signal at 9.70–9.79 ppm was 21 and 15% for E-IR and Z-IR, respectively.

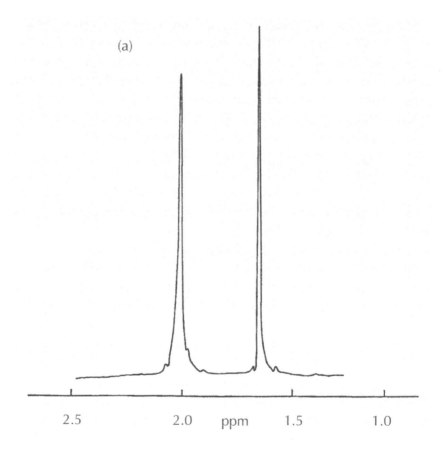

**FIGURE 6**   *(Continued)*

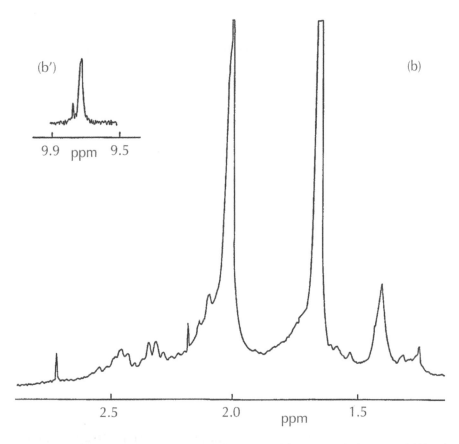

9.9  ppm  9.5

2.5                    2.0                    1.5
                        ppm

**FIGURE 6**  ¹H-250 MHz NMR spectra of 1,4-*cis*-polyisoprene solutions (1 g / 100 ml CCl₄): (a) nonozonized; (b) ozonized to 23% conversion of the double bonds (external standard TMS; digital resolution 0.4 Hz, 20°C).

**TABLE 4**  Assignment of the signals in the ¹H-NMR spectra of partially ozonized 1,4-*cis*-, and 1,4-*trans* polyisoprenes.

| Assignment of the signals (according Fig. 4) | Chemical shifts (ppm) | | Literature |
| --- | --- | --- | --- |
| | **E-IR** | **Z-IR** | |
| a | max 1.41 | max 1.40 | 37, 38 |
| b | 1.70–1.78 | 1.68–1.82 | 37 |
| | | max 1.71 | |
| c | 2.10–2.20 | 2.10–2.20 | 33 |

**TABLE 4**   *(Continued)*

| Assignment of the signals (according Fig. 4) | Chemical shifts (ppm) | | Literature |
|---|---|---|---|
| | **E-IR** | **Z-IR** | |
| d | 2.22–2.40 | 2.22–2.35 | 33 |
| | max 2.33, 2.36 | max 2.29, 2.33 | |
| e | 2.40–2.60 | 2.35–2.55 | 33 |
| | max 2.47 | max 2.48, 2.46 | |
| f | max 2.73 | max 2.73 | 33 |
| g | max 9.76 | max 9.77 | 33 |

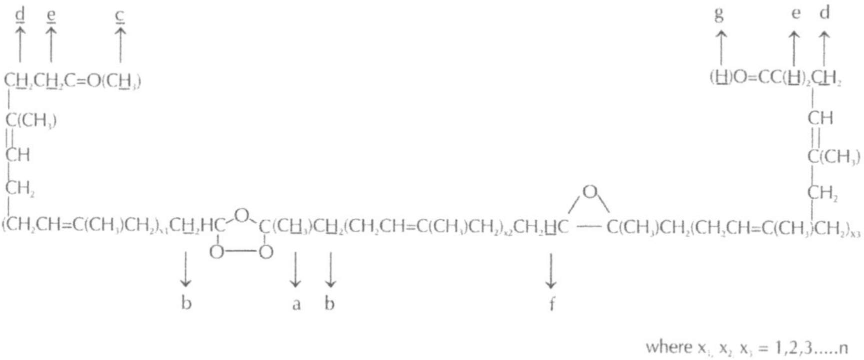

where $x_1, x_2, x_3 = 1,2,3.....n$

**FIGURE 7**   Selection of protons with characteristic signals in the $^1$H-250 MHz NMR spectra of partially ozonized polyisoprene macromolecules.

Current ideas about the mechanism of C=C double bond ozonolysis in solution are summarized in Schemes 1 and 2 [19, 21, 34]. As a result of the decomposition of the initial reaction product, primary ozonide (PO), zwitterionic species is formed, termed as Criegee's intermediates or carbonyl oxides (hereafter referred to as CI) (Scheme 1, reactions 2 and 2'). Two intermediates are formed from asymmetric olefins: monosubstituted CI (MCI) and disubstituted CI (DCI), if their *syn* and *anti* stereoisomers are not taken into account. It is known that carbonyl oxides are predominantly formed at carbon atoms with electron-donating substituents [19]. Excellent correlations of the regioselectivities of MO fragmentation with electron donation by substituents (as measured by Hammett and Taft parameters) have been obtained, consistent with the effects expected for stabilization of a zwiterionic carbonyl oxide [20]. According to ref. [23], for polyisoprenes the ratio between the two intermediates, DCI and MCI, is 64:36.

Ozonides are the basic product of polyisoprene ozonolysis in non-participating solvents. It is known that the dominant part of ozonides is formed through the interaction between CI and the corresponding carbonyl group, which originate from the decomposition of one and the same PO, i.e., a solvent cage effect is operating (Scheme 1, reactions 3 and 3') [19, 34]. With simple olefins the so-called normal ozonides are over 70% [26]. Cross-ozonide formation is also observed (Scheme 1, reactions 5 and 5'). The amount of cross-ozonides is dependent upon the structure of the double bonds, their concentration in the solution, temperature and solvent nature. Reference data indicate that the cross-ozonide yield is strongly reduced at C=C bond concentrations of the order of 0.1 M, with nonpolar solvents, and at temperatures over 0°C [19]. It is reasonable to expect that the polymeric nature of the double bonds in the polyisoprenes would additionally impede the formation of cross- ozonides. Our estimates showed that the amount of cross-ozonides, formed on ozonolysis of both elastomers, is less than 10% of their total quantity. A very small percentage in the overall balance of reacted ozone is the share of the reaction of polymeric ozonides formation (Scheme 1, reactions 4 and 4') [22].

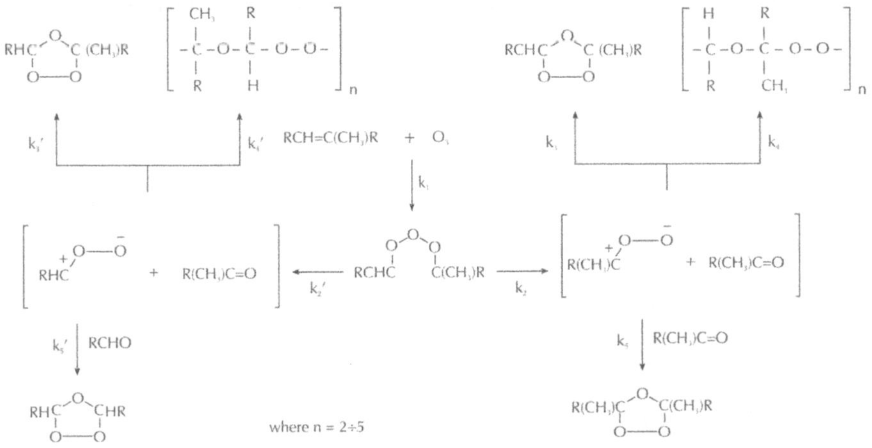

**SCHEME 1**   Criegee's mechanism of formation of ozonides.

In the presence of two types of intermediates, DCI and MCI, it is interesting to follow their further conversion with a view to the reaction products determined by ${}^1$H-NMR spectroscopy. The reaction between CI and carbonyl groups is usually considered to be 1,3-cycloaddition [20, 34]. In this connection the CI-aldehyde interaction is the most effective one: it readily precedes with high ozonide yields. Ketones are considerably less dipolarophilic and good yields of ozonides are limited to special conditions involving particularly reactive ketones, intramolecular reactions, or where ketone is used as the reaction solvent and is Therefore, present in large concentration [19–21]. In order to evaluate the contribution of reactions 3 and 3' (Scheme 1), it is useful to discuss two hypotheses: (i) no reaction proceeds between MCI and ketone, the registered amount of ozonides being formed in reaction 3; (ii) because of the solvent cage effect already mentioned, both the reactions can be considered with practically equal rate constants. Then the ozonide yields would be proportional to the ratio between the two zwitterions (64:36). In this case the corresponding yields of the reactions DCI + aldehyde and MCI + ketone can be calculated from the following system of equations:

$$Y = x + y \tag{6}$$

$$x/y = 64/36 \tag{7}$$

where Y is the total ozonide yield, $x$ and $y$ are the corresponding ozonide yields with the former and the latter reactions, respectively. Results are given in Table 5. A comparison between the aldehyde amounts from both hypotheses and NMR data shows that the former hypothesis leads to results that are much closer to the experimental data.

TABLE 5   Intermediates and products of the ozonolysis of 1,4-*cis*-, and 1,4-*trans* polyisoprenes.

| | DCI + aldehydes (64%) | | | MCI + ketones (36%) | | |
|---|---|---|---|---|---|---|
| | ozonides (%) | aldehydes (%) | DCI (%) | ozonides (%) | ketones (%) | MCI (%) |
| hypothese (i) | | | | | | |
| E – IR | 40 | 24 | 24 | 0 | 36 | 36 |
| Z – IR | 42 | 22 | 22 | 0 | 36 | 36 |
| hypothese (ii) | | | | | | |
| E – IR | 26 | 38 | 38 | 14 | 22 | 22 |
| Z – IR | 27 | 37 | 37 | 15 | 21 | 21 |

More complicated is the question of the other conversion routes with carbonyl oxides outside the solvent cage. It is generally accepted that the deactivation of DCI from low molecular olefins normally takes place via a bimolecular reaction mechanism. At low temperatures ($> -70°C$) dimeric peroxides are formed (Scheme 2, reaction 1) whereas higher temperatures give rise to the formation of carbonyl compounds and evolution of oxygen (Scheme 2, reaction 2) [20]. However, on comparing the yields of keto groups, determined from $^1$H-NMR spectra, with the amounts of ketones and DCI, presented in Table 5, it is seen that reaction 2, if occurring at all, is not dominating during DCI conversion. From kinetic evaluations of CI concentrations in solution it follows that under comparable experimental conditions these concentrations are 4–6 orders of magnitude lower than those of the C=C bonds [23]. Very likely, interactions between two or more DCI are hindered by extremely low values of CI concentration and mostly by the polymeric nature of substituent $R_1$. Under these conditions, DCI tautomerization followed by decomposition of hydroperoxides, formed in the excited state (Scheme 2, reactions 3 and 4), seems most probable.

On going into MCI conversion routes one should have in mind the lower stability and lifetime of these intermediates compared to DCI. Reference data on the proceeding of reactions 1 and 2 (Scheme 2) during ozonolysis of low-molecular olefins in solution are missing [34]. On ozonolysis of 1,4-*trans*-polychloroprene the share of CI of the MCI type is over 80%, in agreement with the induction effect of the chlorine atom [23]. However, no aldehyde groups in the ozonized polymer solution were found [39]. It follows that the interaction between MCI entites, if any, is not the dominating deactivation reaction with these intermediates. The most probable MCI monomolecular deactivation route is assumed to be the isomerization of the MCI intermediates via hot acid to radicals (Scheme 2, reactions 5, 6, 7), because no acid groups were detected (Scheme 2, reaction 8). According to the literature reaction 5 (Scheme 2) does not occur with DCI intermediates [21].

Another route of carbonyl oxide deactivation is double bond epoxidation. Various schemes of olefin epoxidation during ozonolysis have been suggested but the epoxidation via CI (Scheme 2, reaction 9) is presumed to be the most probable with the C=C bonds in polyisoprenes [20, 21]. Since a peak at 2.73 ppm has also been observed under similar conditions on ozonolysis of acrylonitrile-butadiene copolymers [31], as well as during ozonolysis of polybutadienes, it can be assumed that the epoxidation reaction takes place with the participation of both types of CI.

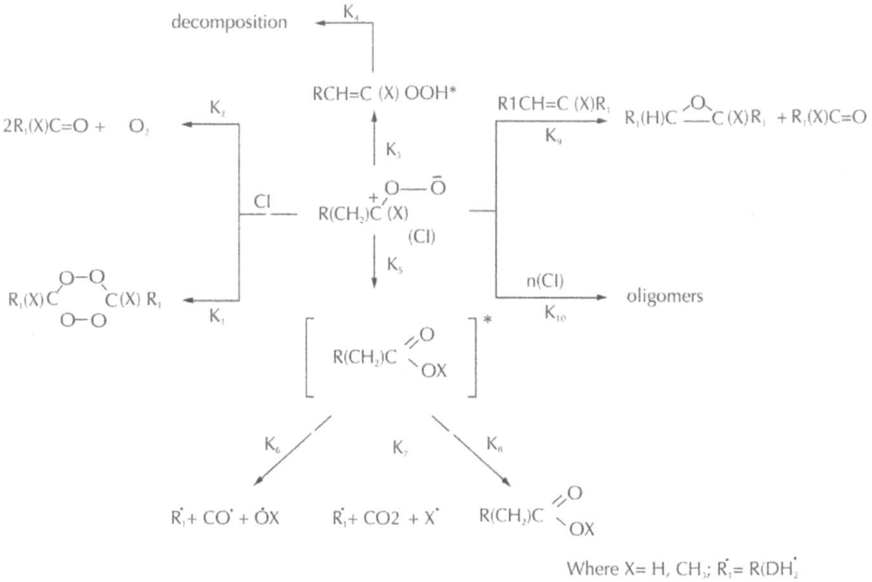

Where X= H, CH₃; Ṙ₁= R(DH)₂

**SCHEME 2**  Non-ozonide routes of deactivation of Criegee's intermediates.

### 1.3.3  POLYCHLOROPRENE

The electron accepting properties of the chlorine atom at the polychloroprene double bond reduces the reactivity of Denka M40 as demonstrated by its relatively low rate constant, i.e., $k=4.10^3$ M$^{-1}$.s$^{-1}$. In this reaction the ratio between the zwitterions A and B, according to theoretical calculations, is in favor of A, the ratio being A/B = 4.55. The A formation is accompanied by choroanhydryde group formation and that of B with aldehyde one. In both cases the ozonides formation is insignificant and the zwitterions react predominantly in the volume resulting in enhancement of the degradation process. The intensive band detected at 1795 cm$^{-1}$ in the IR spectrum of ozonized Denka M40 solutions (Fig. 8) is characteristic of chloroanhydride group [39]. This fact correlated well with the conclusion about the direction of the primary ozonide decomposition. The band at 955

$cm^{-1}$ is also typical of chloroanhydrides. Two other bands – at 1,044 and 905 $cm^{-1}$ may be attributed to the C-O vibrations. The valent vibrations characteristics of HO-groups are observed in the region of 3,050–3,500 $cm^{-1}$.

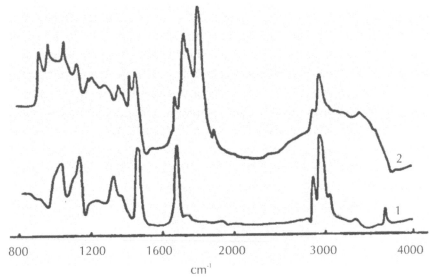

**FIGURE 8**   Infra-red spectra of Denka M40 solutions: 1 – nonozonized; 2 – ozonized to 40% conversion.

The iodometrical analysis of active oxygen in the ozonized Denka M40 solutions shows that the amount of O-O groups is ca. 43%. It is of interest to note that the HI reaction with ozonized polychloroprene solutions occurs quantitatively for 3–4 h, while in SKD the same proceeds only to 20% after 24 h. The above data, however, provide insufficient information for the preferable route of the zwitterions deactivation (*via* dimerization, polymerization of zwitterions or secondary processes). The DSC analysis of the products of Denka M40 ozonolyis reveals that the chloroprene rubber ozonolyis yields polyperoxide as the enthalpy of its decomposition is found to be very close to that of dicumeneperoxide (DCP), The higher value of $E_a$ (ca. two times of that of DCP) testifies the possible formation of polymer peroxides [4].

## DSC STUDY OF THERMAL DECOMPOSITION OF PARTIALLY OZONIZED DIENE RUBBERS

The thermal decomposition of the ozonized diene rubbers can exercise influence on the aging processes as an initiator of the oxidation reactions [40]. On the other hand it is of interest for the elastomers modification and oligomerization [6]. The importance of the DSC method to the investigation of 1,2,4-trioxolanes is based on high values of the enthalpy of the reaction of ozonide thermal decomposition and on the temperature range in which it takes place [41–43]. An intense and relatively broad exothermic peak is characteristic for the thermograms of the partially ozonated 1,4-*cis*-polybutadiene (E-BR), Diene 35 NFA (BR), 1,4-*cis*-polyisoprene (E-IR), 1,4-*trans*-polyisoprene (Z-IR) and 1,4-*trans*-polychloroprene (PCh), recorded in 60–200°C range (Fig 9). Practically no thermal effects are detected in the respective thermograms of the nonozonized samples.

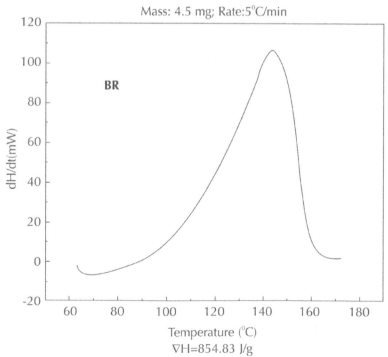

**FIGURE 9**   DSC curve of partially ozonated Diene 35 NFA rubber (BR).

For an objective consideration on the enthalpy changes it is necessary to normalize DH values with respect to the amount of consumed ozone (G). It is distributed on the whole mass of the ozonated sample. The whole mass is a sum of the mass of nonozonated rubber and mass of the incorporated in diene macromolecules oxygen atoms, grouped as a moles ozone. Determination of the amounts of the incorporated ozone ($G_{inc}$) in rubber samples is performed in [43]. The corresponding values of the coefficient of incorporation of ozone ($c_{inc}$), defined as a ratio of the amounts of incorporated ($G_{inc}$) and consumed (G) ozone is presented in Table 6. Using $DH_1$ values and ratio of the functional group, deduced from the NMR spectra, the enthalpy of the ozonide thermal decomposition ($DH_2$) has been evaluated. The data of Table 6 clearly show tendency of increasing of $DH_2$ values with the number of alkyl substituents of the ozonides.

TABLE 6    Enthalpy changes on the thermolysis of partially ozonated diene rubbers.

| Sample | $\Delta H$, (J/g) | $c_{inc}$ | $\Delta H_1$, (kJ/mole $O_3$) | Ozonide (mole)/Incorp. ozone (mole) | $\Delta H_2$ (kJ/mole ozonide) |
|---|---|---|---|---|---|
| E-BR | 954.07 | 0.93 | 332 | 0.96 | 373 |
| BR | 854.83 | 0.87 | 271 | 0.89 | 350 |
| E-IR | 576.19 | 0.62 | 182 | 0.65 | 453 |
| Z-IR | 576.62 | 0.62 | 178 | 0.67 | 426 |
| PCh | 704.57 | 0.81 | 254 | | |
| 1-decene ozonide | | | | | 349 [7] |

There are no considerable differences between the values of the activation energy (E) and reaction order (n) of the ozonide thermal decomposition with E-IR and Z-IR (Table 7). The smaller E values of the polyisoprene ozonides in comparison with those of E-BR ozonides and 1-decene ozonide are, most probably, due to the lower thermal stability of small amounts of oligomeric peroxides, which are present among the reaction products of E-IR and Z-IR ozonolysis [36].

**TABLE 7**    DSC analysis of the ozonized rubbers decomposition [43].

|                  | *$T_m$ (°C) | $a_s$ | E (kJ/mole) | n    |
|------------------|-------------|-------|-------------|------|
| E-BR             | 147         | 0.67  | 125         | 1.0  |
| BR               | 145         | 0.68  | 117         | 0.96 |
| E-IR             | 160         | 0.68  | 111         | 1.04 |
| Z-IR             | 149         | 0.68  | 113         | 1.08 |
| PCh              | 96          | 0.55  | 70          | 1.25 |
| 1-decene ozonide | 117         | 0.63  | 132         | 1.0  |

*$T_m$ and $a_s$ values at heating rate of 5°C/min; $T_m$ and $a_s$ are temperature and degree of conversion at the DSC peak maximum.

## 1.4  CONCLUSIONS

The ozone reaction with a number of polyidienes with different configurations of the double bonds and various substituents was investigated in $CCl_4$ solution.

The changes of the viscosity of the polymer solutions during the ozonolysis were characterized by the determination of the number of chain scissions per molecule of reacted ozone (j). The influence of the conditions of mass-transfer of the reagents in a bubble reactor on the respective j values were discussed.

The basic functional groups-products from the rubbers ozonolysis were identified and quantitatively characterized by means of IR-spectroscopy and $^1$H-NMR spectroscopy. The aldehyde:ozonide ratio was 11:89 and 27:73 for E-BR and BR, respectively. In addition, epoxide groups were detected, only in the case of BR, their yield was about 10% of that of the aldehydes. On polyisoprenes the ozonide:ketone:aldehyde ratio, was 40:37:23 and 42:39:19 for E-IR and Z-IR, respectively. Besides the already specified functional groups, epoxide groups were also detected, their yields being 8 and 7% for E-IR and Z-IR, respectively, with respect to reacted ozone. In the case of 1,4-*trans*-polychloroprene the chloroanhydride group was found to be the basic carbonyl product.

A reaction mechanism, that explains the formation of all identified functional groups, was proposed. It has been shown that the basic route of the reaction of ozone with elastomer double bonds – the formation of normal ozonides does not lead directly to a decrease in the molecular mass of the elastomer macromolecules, because the respective 1,2,4-trioxolanes are relatively stable at ambient temperature. The most favorable conditions for ozone degradation emerge when the cage interaction between Crigee's intermediates and respective carbonyl groups does not proceed. The amounts of measured different carbonyl groups have been used as a alternative way for evaluation of the intensity and efficiency of the ozone degradation.

The thermal decomposition of partially ozonated diene rubbers was investigated by DSC. The respective values of the enthalpy, the activation energy and the reaction order of the 1,2,4-trioxolanes were determined.

## KEYWORDS

- **anomalous products**
- **¹H-NMR spectroscopy**
- **IR-spectroscopy**
- **reacted ozone**

## REFERENCES

1. Zuev, Yu. S. In: *Degradation of Polymers by Action of Aggressive Medias*, Khimia Publishing House: Moscow, Russia, **1972**.
2. Razumovskii, S. D.; Zaikov, G. E. In: *Developments in Polymer Stabilization*, Volume 6, Ed. Scott, G., Applied Science Publishing House: London, **1983**.
3. Jellinek, H. H. G. In: *Aspects of Degradation and Stabilization of Polymers*, Elsevier: Amsterdam, Nederland, **1978,** Chapter 9.
4. Zaikov, G. E.; Rakovsky, S. K. In: *Ozonation of Organic and Polymer Compounds*, Smithres Rapra, UK, **2009,** Chapters 3 and 4.
5. Sweenew, G. P.; Ozonolysis of natural rubber, *J. Polym. Sci.; Part A-1*, 6, **1968,** 26–79.
6. Egorova, G. G.; Shagov, S., "Ozonolysis in the Chemistry of Unsaturated Polymers", in: *synthesis and Chemical Transformation of Polymers*, Izd. LGU, Leningrad, **1986** (In Russian).

7.  Hackathorn, M. J.; Brock, M. J. The determination of "Head-head" and "Tail-tail" structure in polyisoprene, *Rubb. Chem. Technol. 45(6),* **1972,** 12–95.
8.  Hackathorn, M. J.; Brock, M. J.; Altering structures in copolymers as elucidated by microoozonolysis, *J. Polym. Sci.; Part A-1,* **1968,** *6,* 945.
9.  Beresnev, V. V.; Grigoriev, E. I. Influence of the ozonation conditions on molecular mass distribution of oligodienes with end functional groups, *Kautchuk i Rezina,* **1993,** *4,* 10.
10. Nor, H. M.; Ebdon, J. R. Telechelic, liquid natural rubber: A review, *Prog. Polym. Sci.;* **1998,** *23,* 143.
11. Solanky, S. S.; Singh, R. P. Ozonolysis of Natural Rubber: A Critical Review, *Prog. Rubb. Plastics. Technol.;* **2001,** *17(1),* 13.
12. Robin, J. J. The Use of Ozone in the Synthesis of New Polymers and the Modification of Polymers, *Adv. Polym. Sci,* **2004,** *167,* 35.
13. Murray, R.; Story, P. Ozonation, in *Chemical Reactions of Polymers,* Chapter 3, Mir, Moscow, **1967,**
14. Cataldo, F. The action of ozone on polymers having unconjugated and cross- or linearly conjugated unsaturation: chemistry and technological aspects, *Polym. Deg. Stab.* **2001,** *73,* 511.
15. Montaudo, G.; Scamporrino, E.; Vitalini, D. Rapisardi, R. Fast Atom Bombardment Mass Spectrometric Analysis of the Partial Ozonolysis Products of Poly(isoprene) and Popy(chloroprene), *Polym. J., Sci.:* Part A, **1992,** *30,* 525.
16. Ivan, G.; Giurginca, M. Ozone destruction of some transpolydienes, *Polym. Deg. Stab.;* **1998,** *62,* 441.
17. Nor, H. M.; Ebdon, J. R. Ozonolisys of natural rubber in chloroform solution Patr 1. A study by GPC and FTIR spectroscopy, *Polymer,* **2000,** *41,* 23–59.
18. Somers, A. E.; Bastow, T. J.; Burgar, M. I.; Forsyth, M.; Hill, A. J. Quantifying rubber degradation using NMR, *Polym. Deg. Stab.;* **2000,** *70,* 31.
19. Bailey, P. S. *Ozonation in organic chemistry,* Vols. *1, 2,* Academic Press, New York **1978, 1982.**
20. Bunnelle, W. H. Preparation, Properties, and Reactions of Carbonyl Oxides, *Chem. Rev.;* **1991,** *91,* 335.
21. McCullough, K. J.; Nojima, M. "Peroxides from Ozonization", In: *Organic Peroxides,* Ando W.; Chapter *13,* Ed.; John Wiley and Sons Ltd, **1992.**
22. Anachkov, M. P.; Rakovsky, S. K.; Shopov, D. M.; Razumovskii, S. D.; Kefely, A. A.; Zaikov, G. E. Study of the ozone degradation of polybutadiene, polyisoprene and polychloroprene in solution, *Polym. Deg. Stab.;* **1985,** *10(1),* 25.
23. Razumovskii, S. D.; Rakovski, S. K.; Shopov, D. M.; Zaikov, G. E.; *Ozone and Its Reactions with Organic Compounds (in Russian).* Publishing House of Bulgarian Academy of Sciences, Sofia, **1983.**
24. Flory, P. J. *Principles of Polymer Chemistry.* New York, USA, **1953,** *41.*
25. Rakovsky, S.; Zaikov, G. Ozonolysis of polydienes, *Appl. J. Polym. Sci.;* **2004,** *91(3),* 2048.
26. Berlin, A. A.; Sayadyan, A. A.; Enikolopyan, N. S. Molecular-weight distribution and reactivity of polymers in concentrated and diluted solutions, *Visoko Molekul. Soedin.* – A, **1969,** *11,* 1893.

27. Razumovskii, S. D.; Niazashvili, G. A.; Yur'ev, Yu. N.; Tutorskii, I. A. Synthesis and study of polymeric ozonides, *Visoko Molekul. Soedin.* – A, **1971,** *13,* 195.
28. Razumovskii, S. D.; Zaikov, G. E. Kinetics and mechanism of ozone reaction with double bonds, *Uspekhi khimii,* **1980,** *49,* 2344.
29. Kefely, A. A.; Razumovskii, S. D.; Zaikov, G. E. Reaction of ozone with polyethylene, *Visoko Molekul Soedin.* – A, **1971,** *13,* 803.
30. Rakovsky, S. K.; Cherneva, D. R.; Shopov, D. M.; Razumovskii, S. D. Applying of Barbotage Method to Investigation the Kinetic of Ozone Reactions with Organic Compounds.; *Izv. Khim. BAN,* **1976,** *9(4),* 711.
31. Anachkov, M. P.; Rakovsky, S. K.; Stefanova, R. Shopov, D. M. Kinetics and mechanism of the ozone degradation of nitrile rubbers in solution, *Polym. Deg. Stab.;* **1987,** *19(2),* 293.
32. Anachkov M. P.; Rakovsky, S. K.; Zaikov, G. E. Ozonolysis of polybutadienes with different microstructure in solution, *Appl J, Polym. Sci.;* **2007,** *104(1),* 427.
33. *The Sadtler Handbook of Proton NMR Spectra,* Sadtler, Philadelphia, PA, **1978.**
34. Kuczkowski, R. L.; The Structure and Mechanism of Formation of Ozonides, *Chem. Soc. Reviews,* **1992,** *21,* 79.
35. Nakanishi, K. *Infrared spectra and the structure of organic compounds,* Mir, Moscow, **1965.**
36. Anachkov, M. P.; Rakovsky, S. K.; Stefanova, R. V. Ozonolysis of 1, 4-*cis*-polyisoprene and 1, 4-trans-polyisoprene in solution, *Polym. Deg. Stab,* **2000,** *67(2),* 355.
37. Murray, R. W.; Kong, W.; Rajadhyaksha, S. N. The ozonolysis of tetra-methylethylene – concentration and temperature effects *Org. J. Chem.;* **1993,** *58,* 315.
38. Choe, J.-I.; Sprinivasan, M.; Kuczkowski, R. L. Mechanism of the Ozonolysis of Propene in the Liquid-Phase *Am. J. Chem. Soc.;* **1983,** *105,* 4703.
39. Anachkov, M. P.; Rakovsky, S. K.; Stefanova, R. V.; Stoyanov, A. K. Ozone Degradation of Polychloroprene Rubber in Solution, *Polym. Deg. Stab.,* **1993,** *41(2),* 185.
40. Fotty, R. K.; Rakovsky, S. K.; Anachkov, M. P.; Ivanov, S. K. Cumene Oxidation in the Presence of 1-Decene-Ozonide, *Oxidation Commun.;* **1997,** *20(3),* 411.
41. Anachkov, M. P.; Rakovsky, S. K.; Stoyanov, A. K.; Fotty, R. K. DSC Study of the Thermal Decomposition of 1-decene Ozonide, *Thermochim. Acta,* **1994,** *237,* 213.
42. Anachkov, M. P.; Rakovsky, S. K.; Stoyanov, A. K.; DSC Study of the Thermal Decomposition of Partially Ozonized Diene Rubbers, *Appl. J. Polym. Sci.;* **1996,** *61,* 585.
43. Anachkov, M. P.; Rakovsky, S. K. Enthalphy of the Thermal Decomposition of Functional Groups of Peroxide Type, Formed by Ozonolysis of Diene Rubbers in Solution, *Bulg. Chem. Comm.;* **2002,** *34(3/4),* 486.

# CHAPTER 2

# ON POLYMER NANOCOMPOSITES STRUCTURE

G. V. KOZLOV, YU. G. YANOVSKII, E. M. PEARCE, and G. E. ZAIKOV

## CONTENTS

## 2.1   INTRODUCTION

The experimental analysis of particulate-filled nanocomposites butadiene-stirene rubber/fullerene-containing mineral (nanoshungite) was fulfilled with the aid of force-atomic microscopy, nanoindentation methods and computer treatment. The theoretical analysis was carried out within the frameworks of fractal analysis. It has been shown that interfacial regions in the mentioned nanocomposites are the same reinforcing element as nanofiller actually. The conditions of the transition from nano- to microsystems were discussed. The fractal analysis of nanoshungite particles aggregation in polymer matrix was performed. In has been shown that reinforcement of the studied nanocomposites is a true nanoeffect.

The modern methods of experimental and theoretical analysis of polymer materials structure and properties allow not only to confirm earlier propounded hypotheses, but to obtain principally new results. Let us consider some important problems of particulate-filled polymer nanocomposites, the solution of which allows to advance substantially in these materials properties understanding and prediction. Polymer nanocomposites multicomponentness (multiphaseness) requires their structural components quantitative characteristics determination. In this aspect interfacial regions play a particular role, since it has been shown earlier, that they are the same reinforcing element in elastomeric nanocomposites as nanofiller actually [1]. Therefore, the knowledge of interfacial layer dimensional characteristics is necessary for quantitative determination of one of the most important parameters of polymer composites in general – their reinforcement degree [2, 3].

The aggregation of the initial nanofiller powder particles in more or less large particles aggregates always occurs in the course of technological process of making particulate-filled polymer composites in general [4] and elastomeric nanocomposites in particular [5]. The aggregation process tells on composites (nanocomposites) macroscopic properties [2–4]. For nanocomposites nanofiller aggregation process gains special significance, since its intensity can be the one, that nanofiller particles aggregates size exceeds 100 nm – the value, which is assumed (though conditionally enough [6]) as an upper dimensional limit for nanoparticle. In other words, the aggregation process can result to the situation when primordially supposed nanocomposite ceases to be one. Therefore, at present several methods exist, which allow to suppress nanoparticles aggregation process [5, 7]. This also assumes the necessity of the nanoparticles aggregation process quantitative analysis.

It is well-known [1, 2], that in particulate-filled elastomeric nanocomposites (rubbers) nanofiller particles form linear spatial structures ("chains"). At the same time in polymer composites, filled with disperse microparticles (microcomposites) particles (aggregates of particles) of filler form a fractal network, which defines polymer matrix structure (analog of fractal lattice in computer simulation) [4]. This results to different mechanisms of polymer matrix structure formation in micro and nanocomposites. If in the first filler particles (aggregates of particles) fractal network availability results to "disturbance" of polymer matrix structure, that is expressed in the increase of its fractal dimension $d_f$ [4], then in case of polymer nanocomposites at nanofiller contents change the value $d_f$ is not changed and equal to matrix polymer structure fractal dimension [3]. As it has been expected, the change of the composites of the indicated classes structure formation mechanism change defines their properties, in particular, reinforcement degree [11, 12]. Therefore, nanofiller structure fractality strict proof and its dimension determination are necessary.

As it is known [13, 14], the scale effects in general are often found at different materials mechanical properties study. The dependence of failure stress on grain size for metals (Holl-petsch formula) [15] or of effective filling degree on filler particles size in case of polymer composites [16] are examples of such effect. The strong dependence of elasticity modulus on nanofiller particles diameter is observed for particulate-filled elastomeric nanocomposites [5]. Therefore, it is necessary to elucidate the physical grounds of nano- and micromechanical behavior scale effect for polymer nanocomposites.

At present a disperse material wide list is known, which is able to strengthen elastomeric polymer materials [5]. These materials are very diverse on their surface chemical constitution, but particles small size is a common feature for them. On the basis of this observation the hypothesis was offered, that any solid material would strengthen the rubber at the condition, which it was in a very dispersed state and it could be dispersed in polymer matrix. Edwards [5] points out, that filler particles small size is necessary and, probably, the main requirement for reinforcement effect realization in rubbers. Using modern terminology, one can say, that for rubbers reinforcement the nanofiller particles, for which their aggregation process is suppressed as far as possible, would be the most effective ones [3, 12]. Therefore, the theoretical analysis of a nanofiller particles size influence on polymer nanocomposites reinforcement is necessary.

Proceeding from the said above, the present work purpose is the solution of the considered above paramount problems with the help of modern experimental and theoretical techniques on the example of particulate-filled butadiene-stirene rubber.

## 2.2   EXPERIMENTAL METHODS

The made industrially butadiene-stirene rubber of mark SKS-30, which contains 7.0–12.3% cis- and 71.8–72.0% trans-bonds, with density of 920–930 $kg/m^3$ was used as matrix polymer. This rubber is fully amorphous one.

Fullerene-containing mineral shungite of Zazhoginsk's deposit consists of ~30% globular amorphous metastable carbon and ~70% high-disperse silicate particles. Besides, industrially made technical carbon of mark № 220 was used as nanofiller. The technical carbon, nano- and microshugite particles average size makes up 20, 40 and 200 nm, respectively. The indicated filler content is equal to 37 mass%. Nano- and microdimensional disperse shungite particles were prepared from industrially output material by the original technology processing. The size and polydispersity analysis of the received in milling process shungite particles was monitored with the aid of analytical disk centrifuge (CPS Instruments, Inc., USA), allowing to determine with high precision size and distribution by the sizes within the range from 2 nm up to 50 mcm.

Nanostructure was studied on atomic-forced microscopes Nano-DST (Pacific Nanotechnology, USA) and Easy Scan DFM (Nanosurf, Switzerland) by semicontact method in the force modulation regime. Atomic-force microscopy results were processed with the help of specialized software package SPIP (Scanning Probe Image Processor, Denmark). SPIP is a powerful programs package for processing of images, obtained on

SPM, AFM, STM, scanning electron microscopes, transmission electron microscopes, interferometers, confocal microscopes, profilometers, optical microscopes and so on. The given package possesses the whole functions number, which are necessary at images precise analysis, in a number of which the following ones are included:

- the possibility of three-dimensional reflecting objects obtaining, distortions automatized leveling, including Z-error mistakes removal for examination of separate elements and so on;
- quantitative analysis of particles or grains, more than 40 parameters can be calculated for each found particle or pore: area, perimeter, mean diameter, the ratio of linear sizes of grain width to its height distance between grains, coordinates of grain center of mass a.a. can be presented in a diagram form or in a histogram form.

The tests on elastomeric nanocomposites nanomechanical properties were carried out by a nanointentation method [17] on apparatus Nano Test 600 (Micro Materials, Great Britain) in loades wide range from 0.01 mN up to 2.0 mN. Sample indentation was conducted in 10 points with interval of 30 mcm. The load was increased with constant rate up to the greatest given load reaching (for the rate 0.05 mN/s⁻1 mN). The indentation rate was changed in conformity with the greatest load value counting, that loading cycle should take 20 s. The unloading was conducted with the same rate as loading. In the given experiment the "Berkovich indentor" was used with the angle at the top of 65.3° and rounding radius of 200 nm. Indentations were carried out in the checked load regime with preload of 0.001 mN.

For elasticity modulus calculation the obtained in the experiment by nanoindentation course dependences of load on indentation depth (strain) in ten points for each sample at loads of 0.01, 0.02, 0.03, 0.05, 0.10, 0.50, 1.0 and 2.0 mN were processed according to Oliver-pharr method [18].

## 2.3   RESULTS AND DISCUSSION

In Fig. 1 the obtained according to the original methodics results of elasticity moduli calculation for nanocomposite butadiene-stirene rubber/ nanoshungite components (matrix, nanofiller particle and interfacial layers), received in interpolation process of nanoindentation data, are presented. The processed in SPIP polymer nanocomposite image with shungite nanoparticles allows experimental determination of interfacial layer thickness $l_{if}$, which is presented in Fig. 1 as steps on elastomeric matrix-nanofiller boundary. The measurements of 34 such steps (interfacial layers) width on the processed in SPIP images of interfacial layer various section gave the mean experimental value $l_{if}$=8.7 nm. Besides, nanoindentation results (Fig. 1, figures on the right) showed, that interfacial layers elasticity modulus was only by 23–45% lower than nanofiller elasticity modulus, but it was higher than the corresponding parameter of polymer matrix in 6.0–8.5 times. These experimental data confirm that for the studied nanocomposite interfacial layer is a reinforcing element to the same extent, as nanofiller actually [1, 3, 12].

Let us fulfill further the value $l_{if}$ theoretical estimation according to the two methods and compare these results with the ones obtained experimentally. The first method simulates interfacial layer in polymer composites as a result of interaction of two fractals – polymer matrix and nanofiller surface [19, 20]. In this case there is a sole linear scale $l$, which defines these fractals interpenetration distance [21]. Since nanofiller elasticity modulus is essentially higher, than the corresponding parameter for rubber (in the considered case – in 11 times, see Fig. 1), then the indicated interaction reduces to nanofiller indentation in polymer matrix and then $l=l_{if}$. In this case it can be written [21]:

$$l_{if} \approx a\left(\frac{R_p}{a}\right)^{2(d-d_{surf})/d}$$  (1)

where $a$ is a lower linear scale of fractal behavior, which is accepted for polymers as equal to statistical segment length $l_{st}$ [22], $R_p$ is a nanofiller particle (more precizely, particles aggregates) radius, which for nanoshungite is equal to ~84 nm [23], $d$ is dimension of Euclidean space, in which fractal is considered (it is obvious, that in our case $d$=3), $d_{surf}$ is fractal dimension of nanofiller particles aggregate surface.

The value $l_{st}$ is determined as follows [24]:

$$l_{st} = l_0 C_\infty$$  (2)

where $l_0$ is the main chain skeletal bond length, which is equal to 0.154 nm for both blocks of butadiene-stirene rubber [25], $C_\ast$ is characteristic ratio, which is a polymer chain statistical flexibility indicator [26], and is determined with the help of the following equation [22]:

$$T_g = 129 \left( \frac{S}{C_\infty} \right)^{1/2}$$
(3)

where $T_g$ is glass transition temperature, equal to 217 K for butadiene-stirene rubber [3], $S$ is macromolecule cross-sectional area, determined for the mentioned rubber according to the additivity rule from the following considerations. As it is known [27], the macromolecule diameter quadrate values are equal: for polybutadiene – 20.7 $Å^2$ and for polystirene – 69.8 $Å^2$. Having calculated cross-sectional area of macromolecule, simulated as a cylinder, for the indicated polymers according to the known geometrical formulas, let us obtain 16.2 and 54.8 $Å^2$, respectively. Further, accepting as $S$ the average value of the adduced above areas, let us obtain for butadiene-stirene rubber $S$=35.5 $Å^2$. Then according to the Eq. (3) at the indicated values $T_g$ and $S$ let us obtain $C_\ast$=12.5 and according to the Eq. (2) – $l_{st}$=1.932 nm.

The fractal dimension of nanofiller surface $d_{surf}$ was determined with the help of the equation [3]:

$$S_u = 410 R_p^{d_{surf} - d}$$
(4)

where $S_u$ is nanoshungite particles specific surface, calculated as follows [28]:

$$S_u = \frac{3}{\rho_n R_p}$$
(5)

where $r_n$ is the nanofiller particles aggregate density, determined according to the formula [3]:

$$\rho_n = 0.188 \left( R_p \right)^{1/3}$$
(6)

The calculation according to the Eqs. (4)–(6) gives $d_{surf}$=2.44. Further, using the calculated by the indicated mode parameters, let us obtain from the Eq. (1) the theoretical value of interfacial layer thickness $l_{if}^T$ =7.8 nm. This value is close enough to the obtained one experimentally (their discrepancy makes up ~10%).

The second method of value $l_{if}^T$ estimation consists in using of the two following equations [3, 29]:

$$\phi_{if} = \phi_n \left( d_{surf} - 2 \right) \qquad (7)$$

and

$$\phi_{if} = \phi_n \left[ \left( \frac{R_p + l_{if}^T}{R_p} \right)^3 - 1 \right] \qquad (8)$$

where $j_{if}$ and $j_n$ are relative volume fractions of interfacial regions and nanofiller, accordingly.

The combination of the indicated equations allows to receive the following formula for $l_{if}^T$ calculation:

$$l_{if}^T = R_p \left[ \left( d_{surf} - 1 \right)^{1/3} - 1 \right] \qquad (9)$$

The calculation according to the Eq. (9) gives for the considered nanocomposite $l_{if}^T$ =10.8 nm, that also corresponds well enough to the experiment (in this case discrepancy between $l_{if}$ and $l_{if}^T$ makes up ~19%).

Let us note in conclusion the important experimental observation, which follows from the processed by program SPIP results of the studied nanocomposite surface scan (Fig. 1). As one can see, at one nanoshungite particle surface from one to three (in average – two) steps can be observed, structurally identified as interfacial layers. It is significant that these steps width (or $l_{if}$) is approximately equal to the first (the closest to nanoparticle surface) step width. Therefore, the indicated observation supposes, that in elastomeric nanocomposites at average two interfacial layers are formed: the first – at the expence of nanofiller particle surface with elastomeric matrix interaction, as a result of which molecular mobility in this layer is frozen and its state is glassy-like one, and the second – at the expence of glassy interfacial layer with elastomeric polymer matrix interaction. The most important question from the practical point of view, whether one interfacial layer or both serve as nanocomposite reinforcing element. Let us fulfill the following quantitative estimation for this question solution. The reinforcement degree ($E_n/E_m$) of polymer nanocomposites is given by the equation [3]:

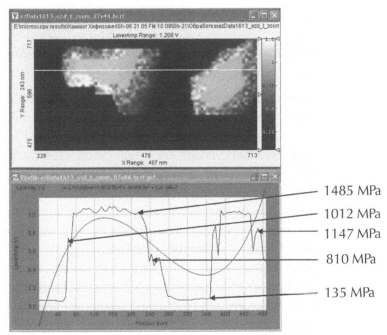

FIGURE 1   The processed in SPIP image of nanocomposite butadiene-stirene rubber/ nanoshungite, obtained by force modulation method, and mechanical characteristics of structural components according to the data of nanoindentation (strain 150 nm).

$$\frac{E_n}{E_m} = 1 + 11\left(\phi_n + \phi_{if}\right)^{1.7} \tag{10}$$

where $E_n$ and $E_m$ are elasticity moduli of nanocomposite and matrix polymer, accordingly ($E_m$=1.82 MPa [3]).

According to the Eq. (7) the sum ($j_n$+$j_{if}$) is equal to:

$$\phi_n + \phi_{if} = \phi_n\left(d_{surf} - 1\right) \tag{11}$$

if one interfacial layer (the closest to nanoshungite surface) is a reinforcing element and

$$\phi_n + 2\phi_{if} = \phi_n \left( 2d_{surf} - 3 \right) \qquad (12)$$

if both interfacial layers are a reinforcing element.

In its turn, the value $j_n$ is determined according to the equation [30]:

$$\phi_n = \frac{W_n}{\rho_n} \qquad (13)$$

where $fi_n$ is nanofiller mass content, $r_n$ is its density, determined according to the Eq. (6).

The calculation according to the Eqs. (11) and (12) gave the following $E_n/E_m$ values: 4.60 and 6.65, respectively. Since the experimental value $E_n/E_m$=6.10 is closer to the value, calculated according to the Eq. (12), then this means that both interfacial layers are a reinforcing element for the studied nanocomposites. Therefore, the coefficient 2 should be introduced in the equations for value $l_{if}$ determination (for example, in the Eq. (1)) in case of nanocomposites with elastomeric matrix. Let us remind, that the Eq. (1) in its initial form was obtained as a relationship with proportionality sign, i.e., without fixed proportionality coefficient [21].

Thus, the used above nanoscopic methodics allow to estimate both interfacial layer structural special features in polymer nanocomposites and its sizes and properties. For the first time it has been shown, that in elastomeric particulate-filled nanocomposites two consecutive interfacial layers are formed, which are a reinforcing element for the indicated nanocomposites. The proposed theoretical methodics of interfacial layer thickness estimation, elaborated within the frameworks of fractal analysis, give well enough correspondence to the experiment.

For theoretical treatment of nanofiller particles aggregate growth processes and final sizes traditional irreversible aggregation models are inapplicable, since it is obvious, that in nanocomposites aggregates a large number of simultaneous growth takes place. Therefore, the model of multiple growth, offered in paper [6], was used for nanofiller aggregation description.

In Fig. 2 the images of the studied nanocomposites, obtained in the force modulation regime, and corresponding to them nanoparticles aggregates fractal dimension $d_f$ distributions are adduced. As it follows from the adduced values $d_f^{ag}$ ($d_f^{ag}$ = 2.40–2.48), nanofiller particles aggregates in the studied nanocomposites are formed by a mechanism particle-cluster (P-cl), i.e., they are Witten-sander clusters [32]. The variant A, was chosen which according to mobile particles are added to the lattice, consisting of a large number of "seeds" with density of $c_0$ at simulation beginning [31]. Such model generates the structures, which have fractal geometry on length short scales with value $d_f \gg 2.5$ (see Fig. 2) and homogeneous structure on length large scales. A relatively high particles concentration $c$ is required in the model for uninterrupted network formation [31].

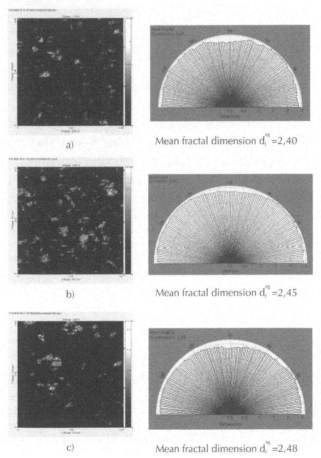

a)

Mean fractal dimension $d_f^{ag}$ =2,40

b)

Mean fractal dimension $d_f^{ag}$ =2,45

c)

Mean fractal dimension $d_f^{ag}$ =2,48

**FIGURE 2** The images obtained in the force modulation regime for nanocomposites are filled with technical carbon (a), nanoshungite (b), microshungite (c) and corresponding to them fractal dimensions $d_f^{ag}$.

In case of "seeds" high concentration $c_0$ for the variant A the following relationship was obtained [31]:

$$R_{max}^{d_f^{ag}} = N = c/c_0 \qquad (14)$$

where $R_{max}$ is nanoparticles cluster (aggregate) greatest radius, $N$ is nanoparticles number per one aggregate, $c$ is nanoparticles concentration, $c_0$ is "seeds" number, which is equal to nanoparticles clusters (aggregates) number.

The value $N$ can be estimated according to the following equation [8]:

$$2R_{max} = \left(\frac{S_n N}{\pi \eta}\right)^{1/2} \qquad (15)$$

where $S_n$ is cross-sectional area of nanoparticles, of which an aggregate consists, h is a packing coefficient, equal to 0.74 [28].

The experimentally obtained nanoparticles aggregate diameter $2R_{ag}$ was accepted as $2R_{max}$ (Table 1) and the value $S_n$ was also calculated according to the experimental values of nanoparticles radius $r_n$ (Table 1). In Table 1 the values $N$ for the studied nanofillers, obtained according to the indicated method, were adduced. It is significant that the value $N$ is a maximum one for nanoshungite despite larger values $r_n$ in comparison with technical carbon.

TABLE 1   The parameters of irreversible aggregation model of nanofiller particles aggregates growth.

| Nanofiller | $R_{ag}$, nm | $r_n$, nm | $N$ | $R_{max}^T$, nm | $R_{ag}^T$, nm | $R_c$, nm |
|---|---|---|---|---|---|---|
| Technical carbon | 34.6 | 10 | 35.4 | 34.7 | 34.7 | 33.9 |
| Nanoshungite | 83.6 | 20 | 51.8 | 45.0 | 90.0 | 71.0 |
| Microshungite | 117.1 | 100 | 4.1 | 15.8 | 158.0 | 255.0 |

Further the Eq. (14) allows to estimate the greatest radius $R_{max}^T$ of nanoparticles aggregate within the frameworks of the aggregation model [31]. These values $R_{max}^T$ are adduced in Table 1, from which their reduction in a sequence of technical carbon-nanoshungite-microshungite, that fully contradicts to the experimental data, i.e., to $R_{ag}$ change (Table 1). However, we must not neglect the fact that the Eq. (14) was obtained within the frameworks of computer simulation, where the initial aggregating particles sizes are the same in all cases [31]. For real nanocomposites the values $r_n$ can be distinguished essentially (Table 1). It is expected, that the value $R_{ag}$ or $R_{max}^T$ will be the higher, the larger is the radius of nanoparticles, forming aggregate, is i.e., $r_n$. Then theoretical value of nanofiller particles cluster (aggregate) radius $R_{ag}^T$ can be determined as follows:

$$R_{ag}^T = k_n r_n N^{1/d_f^{ag}} \qquad (16)$$

where $k_n$ is proportionality coefficient, in the present work accepted empirically equal to 0.9.

The comparison of experimental $R_{ag}$ and calculated according to the Eq. (16) $R_{ag}^T$ values of the studied nanofillers particles aggregates radius shows their good correspondence (the average discrepancy of $R_{ag}$ and $R_{ag}^T$ makes up 11.4%). Therefore, the theoretical model [31] gives a good correspondence to the experiment only in case of consideration of aggregating particles real characteristics and, in the first place, their size.

Let us consider two more important aspects of nanofiller particles aggregation within the frameworks of the model [31]. Some features of the indicated process are defined by nanoparticles diffusion at nanocomposites processing. Specifically, length scale, connected with diffusible nanoparticle, is correlation length x of diffusion. By definition, the growth phenomena in sites, remote more than x, are statistically independent. Such definition allows to connect the value x with the mean distance between nanofiller particles aggregates $L_n$. The value x can be calculated according to the equation [31]:

$$\xi^2 \approx \tilde{n}^{-1} R_{ag}^{d_f^{ag}-d+2} \qquad (17)$$

where $c$ is nanoparticles concentration, which should be accepted equal to nanofiller volume contents $j_n$, which is calculated according to the Eqs. (6) and (13).

The values $r_n$ and $R_{ag}$ were obtained experimentally (see histogram of Fig. 3). In Fig. 4 the relation between $L_n$ and x is adduced, which, as it is expected, proves to be linear and passing through coordinates origin. This means, that the distance between nanofiller particles aggregates is limited by mean displacement of statistical walks, by which nanoparticles are simulated. The relationship between $L_n$ and x can be expressed analytically as follows:

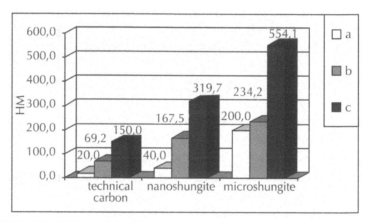

**FIGURE 3**   The initial particles diameter (a), their aggregates size in nanocomposite (b) and distance between nanoparticles aggregates (c) for nanocomposites, filled with technical carbon, nano- and microshungite.

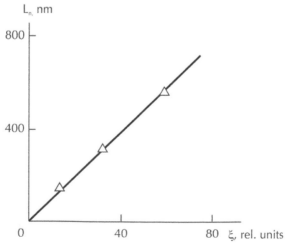

**FIGURE 4**   The relation between diffusion correlation length x and distance between nanoparticles aggregates $L_n$ for considered nanocomposites.

$$L_n \approx 9.6\xi, \text{nm}. \tag{18}$$

The second important aspect of the model [31] in reference to nanofiller particles aggregation simulation is a finite nonzero initial particles concentration $c$ or $j_n$ effect, which takes place in any real systems. This effect is realized at the condition $x \gg R_{ag}$, that occurs at the critical value $R_{ag}(R_c)$, determined according to the relationship [31]:

$$c \sim R_c^{d_f^{ag}-d} \tag{19}$$

The right side of the Eq. (19) represents cluster (particles aggregate) mean density. This equation establishes that fractal growth continues only, until cluster density reduces up to medium density, in which it grows. The calculated according to the Eq. (19) values $R_c$ for the considered nanoparticles are adduced in Table 1, from which follows, that they give reasonable correspondence with this parameter experimental values $R_{ag}$ (the average discrepancy of $R_c$ and $R_{ag}$ makes up 24%).

Since the treatment [31] was obtained within the frameworks of a more general model of diffusion-limited aggregation, then its correspondence to the experimental data indicated unequivocally, that aggregation processes in these systems were controlled by diffusion. Therefore, let us consider briefly nanofiller particles diffusion. Statistical walkers diffusion constant z can be determined with the aid of the relationship [31]:

$$\xi \approx \left(\zeta t\right)^{1/2} \tag{20}$$

where $t$ is walk duration.

The Eq. (20) supposes (at $t$=const) z increase in a number technical carbon-nanoshungite-microshungite as 196–1069–3434 relative units, i.e., diffusion intensification at diffusible particles size growth. At the same time diffusivity $D$ for these particles can be described by the well-known Einstein's relationship [33]:

$$D = \frac{kT}{6\pi\eta r_n\alpha} \tag{21}$$

where $k$ is Boltzmann constant, $t$ is temperature, h is medium viscosity, a is numerical coefficient, which further is accepted equal to 1.

In its turn, the value h can be estimated according to the equation [34]:

$$\frac{\eta}{\eta_0} = 1 + \frac{2.5\phi_n}{1 - \phi_n} \tag{22}$$

where $h_0$ and h are initial polymer and its mixture with nanofiller viscosity, accordingly.

The calculation according to the Eqs. (21) and (22) shows, that within the indicated above nanofillers number the value $D$ changes as 1.32–1.14–0.44 relative units, i.e., reduces in three times, that was expected. This apparent contradiction is due to the choice of the condition $t$=const (where $t$ is nanocomposite production duration) in the Eq. (20). In real conditions the value $t$ is restricted by nanoparticle contact with growing aggregate and then instead of $t$ the value $t/c_0$ should be used, where $c_0$ is the seeds concentration, determined according to the Eq. (14). In this case the value z for the indicated nanofillers changes as 0.288–0.118–0.086, i.e., it reduces in 3.3 times, that corresponds fully to the calculation according to the Einstein's relationship (the Eq. (21)). This means, that nanoparticles diffusion in polymer matrix obeys classical laws of Newtonian rheology [33].

Thus, the disperse nanofiller particles aggregation in elastomeric matrix can be described theoretically within the frameworks of a modified model of irreversible aggregation particle-cluster. The obligatory consideration of nanofiller initial particles size is a feature of the indicated model application to real systems description. The indicated particles diffusion in polymer matrix obeys classical laws of Newtonian liquids hydrodynamics. The offered approach allows to predict nanoparticles aggregates final parameters as a function of the initial particles size, their contents and other factors number.

At present there are several methods of filler structure (distribution) determination in polymer matrix, both experimental [10, 35] and theoretical [4]. All the indicated methods describe this distribution by fractal dimension $D_n$ of filler particles network. However, correct determination of any object fractal (Hausdorff) dimension includes three obligatory conditions. The first from them is the indicated above determination of fractal dimension numerical magnitude, which should not be equal to object topological dimension. As it is known [36], any real (physical) fractal possesses fractal properties within a certain scales range. Therefore, the second condition is the evidence of object self-similarity in this scales range [37]. And at last, the third condition is the correct choice of measurement scales range itself. As it has been shown in papers [38, 39], the minimum range should exceed at any rate one self-similarity iteration.

The first method of dimension $D_n$ experimental determination uses the following fractal relationship [40, 41]:

$$D_n = \frac{\ln N}{\ln \rho} \tag{23}$$

where $N$ is a number of particles with size r.

Particles sizes were established on the basis of atomic-power microscopy data (*see* Fig. 2). For each from the three studied nanocomposites no less than 200 particles were measured, the sizes of which were united into 10 groups and mean values $N$ and r were obtained. The dependences $N(r)$ in double logarithmic coordinates were plotted, which proved to be linear and the values $D_n$ were calculated according to their slope (*see* Fig. 5). It is obvious, that at such approach fractal dimension $D_n$ is determined in two-dimensional Euclidean space, whereas real nanocomposite should be considered in three-dimensional Euclidean space. The following relationship can be used for $D_n$ recalculation for the case of three-dimensional space [42]:

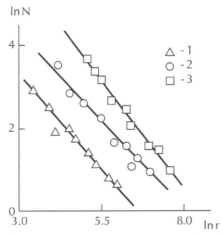

FIGURE 5    The dependences of nanofiller particles number $N$ on their size r for nanocomposites BSR/TC (1), BSR/nanoshungite (2) and BSR/microshungite (3).

$$D3 = \frac{d + D2 \pm \left[(d - D2)^2 - 2\right]^{1/2}}{2} \quad (24)$$

where $D3$ and $D2$ are corresponding fractal dimensions in three- and two-dimensional Euclidean spaces, $d=3$.

The calculated according to the indicated method dimensions $D_n$ are adduced in Table 2. As it follows from the data of this table, the values $D_n$ for the studied nanocomposites are varied within the range of 1.10–1.36, i.e., they characterize more or less branched linear formations ("chains") of nanofiller particles (aggregates of particles) in elastomeric nanocomposite structure. Let us remind that for particulate-filled composites polyhydroxiether/graphite the value $D_n$ changes within the range of ~2.30–2.80 [4, 10], i.e., for these materials filler particles network is a bulk object, but not a linear one [36].

**TABLE 2**  The dimensions of nanofiller particles (aggregates of particles) structure in elastomeric nanocomposites.

| Nanocomposite | $D_n$, the Eq. (23) | $D_n$, the Eq. (25) | $d_0$ | $d_{surf}$ | $\varphi_n$ | $D_n$, the Eq. (29) |
|---|---|---|---|---|---|---|
| BSR/TC | 1.19 | 1.17 | 2.86 | 2.64 | 0.48 | 1.11 |
| BSR/nanoshungite | 1.10 | 1.10 | 2.81 | 2.56 | 0.36 | 0.78 |
| BSR/microshungite | 1.36 | 1.39 | 2.41 | 2.39 | 0.32 | 1.47 |

Another method of $D_n$ experimental determination uses the so-called "quadrates method" [43]. Its essence consists in the following. On the enlarged nanocomposite microphotograph (see Fig. 2) a net of quadrates with quadrate side size $a_i$, changing from 4.5 up to 24 mm with constant ratio $a_{i+1}/a_i=1.5$, is applied and then quadrates number $N_i$, in to which nanofiller particles hit (fully or partly), is counted up. Five arbitrary net positions concerning microphotograph were chosen for each measurement. If nanofiller particles network is a fractal, then the following relationship should be fulfilled [43]:

$$N_i \sim S_i^{-D_n/2} \quad (25)$$

where $S_i$ is quadrate area, which is equal to $\alpha_i^2$.

In Fig. 6 the dependences of $N_i$ on $S_i$ in double logarithmic coordinates for the three studied nanocomposites, corresponding to the Eq. (25), is adduced. As one can see, these dependences are linear, that allows to determine the value $D_n$ from their slope. The determined according to the Eq. (25) values $D_n$ are also adduced in Table 2, from which a good correspondence of dimensions $D_n$, obtained by the two described above methods, follows (their average discrepancy makes up 2.1% after these dimensions recalculation for three-dimensional space according to the Eq. (24)).

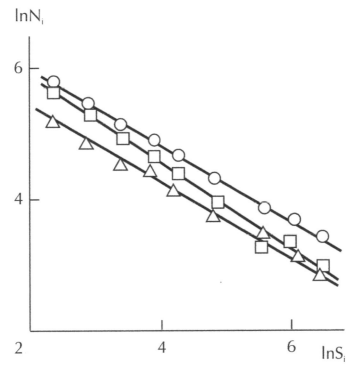

**FIGURE 6** The dependences of covering quadrates number $N_i$ on their area $S_i$, corresponding to the relationship (25), in double logarithmic coordinates for nanocomposites on the basis of BSR. The designations are the same, that in Fig. 5.

As it has been shown in paper [44], the usage for self-similar fractal objects at the Eq. (25) the condition should be fulfilled:

$$N_i - N_{i-1} \sim S_i^{-D_n} \tag{26}$$

In Fig. 7 the dependence, corresponding to the Eq. (26), for the three studied elastomeric nanocomposites is adduced. As one can see, this dependence is linear, passes through coordinates origin, that according to the Eq. (26) is confirmed by nanofiller particles (aggregates of particles) "chains" self-similarity within the selected $a_i$ range. It is obvious, that this self-similarity will be a statistical one [44]. Let us note, that the points, corresponding to $a_i=16$ mm for nanocomposites butadiene-stirene rubber/technical carbon (BSR/TC) and butadiene-stirene rubber/microshungite (BSR/microshungite), do not correspond to a common straight line. Accounting for electron microphotographs of Fig. 2 enlargement this gives the self-similarity range for nanofiller "chains" of 464–1472 nm. For nanocomposite butadiene-stirene rubber/nanoshungite (BSR/nanoshungite), which has no points deviating from a straight line of Fig. 7, $a_i$ range makes up 311–1510 nm, that corresponds well enough to the indicated above self-similarity range.

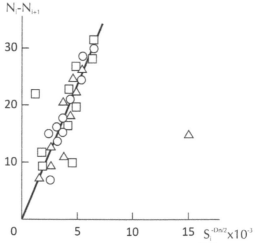

**FIGURE 7** The dependences of $(N_{i-ni+1})$ on the value $S_i^{-D_n/2}$, corresponding to the relationship (26), for nanocomposites on the basis of BSR. The designations are the same, that in Fig. 5.

In papers [38, 39] it has been shown, that measurement scales $S_i$ minimum range should contain at least one self-similarity iteration. In this case the condition for ratio of maximum $S_{max}$ and minimum $S_{min}$ areas of covering quadrates should be fulfilled [39]:

$$\frac{S_{max}}{S_{min}} > 2^{2/D_n} \tag{27}$$

Hence, accounting for the defined above restriction let us obtain $S_{max}/S_{min}=121/20.25=5.975$, that is larger than values $2^{2/D_n}$ for the studied nanocomposites, which are equal to 2.71–3.52. This means, that measurement scales range is chosen correctly.

The self-similarity iterations number m can be estimated from the inequality [39]:

$$\left(\frac{S_{max}}{S_{min}}\right)^{D_n/2} > 2^{\mu} \tag{28}$$

Using the indicated above values of the included in the inequality Eq. (28) parameters, m=1.42–1.75 is obtained for the studied nanocomposites, i.e., in our experiment conditions self-similarity iterations number is larger than unity, that again confirms correctness of the value $D_n$ estimation [35].

And let us consider in conclusion the physical grounds of smaller values $D_n$ for elastomeric nanocomposites in comparison with polymer microcomposites, i.e., the causes of nanofiller particles (aggregates of particles) "chains" formation in the first ones. The value $D_n$ can be determined theoretically according to the equation [4]:

$$\phi_{if} = \frac{D_n + 2.55d_0 - 7.10}{4.18} \tag{29}$$

where $j_{if}$ is interfacial regions relative fraction, $d_0$ is nanofiller initial particles surface dimension.

The dimension $d_0$ estimation can be carried out with the help of the Eq. (4) and the value $j_{if}$ can be calculated according to the Eq. (7). The results of dimension $D_n$ theoretical calculation according to the Eq. (29) are adduced in Table 2, from which a theory and experiment good correspondence follows. The Eq. (29) indicates unequivocally to the cause of a filler in nano- and microcomposites different behavior. The high (close to 3, *see* Table 2) values $d_0$ for nanoparticles and relatively small ($d_0$=2.17 for graphite [4]) values $d_0$ for microparticles at comparable values $j_{if}$ is such cause for composites of the indicated classes [3, 4].

Hence, the stated above results have shown, that nanofiller particles (aggregates of particles) "chains" in elastomeric nanocomposites are physical fractal within self-similarity (and, hence, fractality [41]) range of ~500–1450 nm. In this range their dimension $D_n$ can be estimated according to the Eqs. (23), (25) and (29). The cited examples demonstrate the necessity of the measurement scales range correct choice. As it has been noted earlier [45], the linearity of the plots, corresponding to the Eqs. (23) and (25), and $D_n$ nonintegral value do not guarantee object self-similarity (and, hence, fractality). The nanofiller particles (aggregates of particles) structure low dimensions are due to the initial nanofiller particles surface high fractal dimension.

In Fig. 8 the histogram is adduced, which shows elasticity modulus $E$ change, obtained in nanoindentation tests, as a function of load on indenter P or nanoindentation depth $h$. Since for all the three considered nanocomposites the dependences $E(P)$ or $E(h)$ are identical qualitatively, then further the dependence $E(h)$ for nanocomposite BSR/TC was chosen, which reflects the indicated scale effect quantitative aspect in the most clearest way.

In Fig. 9 the dependence of $E$ on $h_{pl}$ (see Fig. 10) is adduced, which breaks down into two linear parts. Such dependences elasticity modulus – strain are typical for polymer materials in general and are due to intermolecular bonds anharmonicity [46]. In paper [47] it has been shown that the dependence $E(h_{pl})$ first part at $h_{pl} \leq 500$ nm is not connected with relaxation processes and has a purely elastic origin. The elasticity modulus $E$ on this part changes in proportion to $h_{pl}$ as:

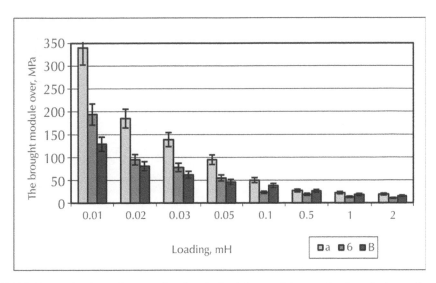

**FIGURE 8** The dependences of reduced elasticity modulus on load on indentor for nanocomposites on the basis of butadiene-stirene rubber, filled with technical carbon (a), micro (b) and nanoshungite (c).

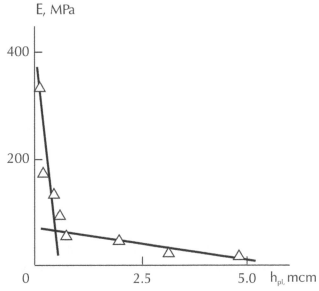

**FIGURE 9** The dependence of reduced elasticity modulus $E$, obtained in nanoindentation experiment, on plastic strain $h_{pl}$ for nanocomposites BSR/TC.

**Berkovich indenter**

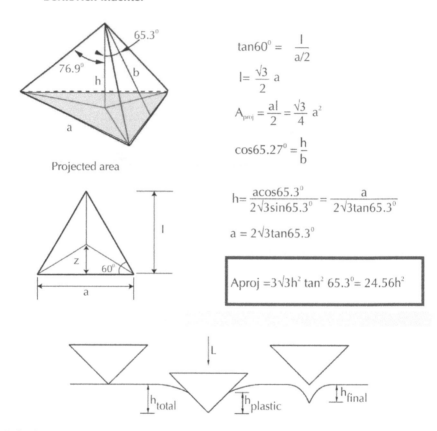

$$\tan 60^\circ = \frac{l}{a/2}$$

$$l = \frac{\sqrt{3}}{2}\,a$$

$$A_{proj} = \frac{al}{2} = \frac{\sqrt{3}}{4}\,a^2$$

$$\cos 65.27^\circ = \frac{h}{b}$$

$$h = \frac{a\cos 65.3^\circ}{2\sqrt{3}\sin 65.3^\circ} = \frac{a}{2\sqrt{3}\tan 65.3^\circ}$$

$$a = 2\sqrt{3}\tan 65.3^\circ$$

$$A_{proj} = 3\sqrt{3}h^2\tan^2 65.3^\circ = 24.56h^2$$

**FIGURE 10**   The schematic image of Berkovich indentor and nanoindentation process.

$$E = E_0 + B_0 h_{pl}, \tag{30}$$

where $E_0$ is "initial" modulus, i.e., modulus, extrapolated to $h_{pl}=0$, and the coefficient $B_0$ is a combination of the first and second kind elastic constants. In the considered case $B_0<0$. Further Grüneizen parameter $g_L$, characterizing intermolecular bonds anharmonicity level, can be determined [47]:

$$\gamma_L \approx -\frac{1}{6} - \frac{1}{2}\frac{B_0}{E_0}\frac{1}{\left(1-2\nu\right)}, \tag{31}$$

where n is Poisson ratio, accepted for elastomeric materials equal to ~0.475 [36].

Calculation according to the Eq. (31) has given the following values $g_L$: 13.6 for the first part and 1.50 – for the second one. Let us note the first from $g_L$ adduced values is typical for intermolecular bonds, whereas the second value $g_L$ is much closer to the corresponding value of Grüneizen parameter $G$ for intrachain modes [46].

Poisson's ratio n can be estimated by $g_L$ (or $G$) known values according to the formula [46]:

$$\gamma_L = 0.7\left(\frac{1+v}{1-2v}\right) \qquad (32)$$

The estimations according to the Eq. (32) gave: for the dependence $E(h_{pl})$ first part n=0.462, for the second one – n=0.216. If for the first part the value n is close to Poisson's ratio magnitude for nonfilled rubber [36], then in the second part case the additional estimation is required. As it is known [48], a polymer composites (nanocomposites) Poisson's ratio value $n_n$ can be estimated according to the equation:

$$\frac{1}{v_n} = \frac{\phi_n}{v_{TC}} + \frac{1-\phi_n}{v_m} \qquad (33)$$

where $j_n$ is nanofiller volume fraction, $n_{TC}$ and $n_m$ are nanofiller (technical carbon) and polymer matrix Poisson's ratio, respectively.

The value $n_m$ is accepted equal to 0.475 [36] and the magnitude $n_{TC}$ is estimated as follows [49]. As it is known [50], the nanoparticles TC aggregates fractal dimension $d_f^{ag}$ value is equal to 2.40 and then the value $n_{TC}$ can be determined according to the equation [50]:

$$d_f^{ag} = (d-1)(1+v_{TC}) \qquad (34)$$

According to the Eq. (34) $n_{TC}$=0.20 and calculation $n_n$ according to the Eq. (33) gives the value 0.283, that is close enough to the value n=0.216 according to the Eq. (32) estimation. The obtained by the indicated methods values n and $n_n$ comparison demonstrates, that in the dependence $E(h_{pl})$

($h_{pl}$<0.5 mcm) the first part in nanoindentation tests only rubber-like polymer matrix (n=$n_m$»0.475) is included and in this dependence the second part – the entire nanocomposite as homogeneous system [51] – n=$n_n$»0.22.

Let us consider further $E$ reduction at $h_{pl}$ growth (Fig. 9) within the frameworks of density fluctuation theory, which value y can be estimated as follows [22]:

$$\psi = \frac{\rho_n kT}{K_T} \qquad (35)$$

where $r_n$ is nanocomposite density, $k$ is Boltzmann constant, $t$ is testing temperature, $K_T$ is isothermal modulus of dilatation, connected with Young's modulus $E$ by the relationship [46]:

$$K_T = \frac{E}{3(1-v)} \qquad (36)$$

In Fig. 10 the scheme of volume of the deformed at nanoindentation material $ff_{def}$ calculation in case of Berkovich indentor using is adduced and in Fig. 11 the dependence y($ff_{def}$) in logarithmic coordinates was shown. As it follows from the data of Fig. 11, the density fluctuation growth is observed at the deformed material volume increase. The plot y($\ln ff_{def}$) extrapolation to y=0 gives $\ln ff_{def}$»13 or $ff_{def}(V_{def}^{cr})$=4.42´$10^5$ nm³. Having determined the linear scale $l_{cr}$ of transition to y=0 as ($V_{def}^{cr}$)$^{1/3}$, let us obtain $l_{cr}$=75.9 nm, that is close to nanosystems dimensional range upper boundary (as it was noted above, conditional enough [6]), which is equal to 100 nm. Thus, the stated above results suppose, that nanosystems are such systems, in which density fluctuations are absent, always taking place in microsystems.

As it follows from the data of Fig. 9, the transition from nano- to microsystems occurs within the range $h_{pl}$=408–726 nm. Both the indicated above values $h_{pl}$ and the corresponding to them values ($ff_{def}$)$^{1/3}$»814–1440 nm can be chosen as the linear length scale $l_n$, corresponding to this transition. From the comparison of these values $l_n$ with the distance between nanofiller particles aggregates $L_n$ ($L_n$=219.2–788.3 nm for the considered nanocomposites, see Fig. 3) it follows, that for transition from nano- to microsystems $l_n$ should include at least two nanofiller particles aggregates and surrounding them layers of polymer matrix, that is the lowest linear scale of nanocomposite simulation as a homogeneous system. It is easy to see, that nanocomposite structure homogeneity condition is harder than the obtained above from the criterion y=0. Let us note, that such method, namely, a nanofiller particle and surrounding it polymer matrix layers separation, is widespread at a relationships derivation in microcomposite models.

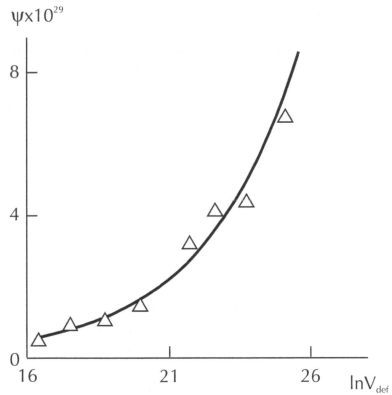

**FIGURE 11** The dependence of density fluctuation y on volume of deformed in nanoindentation process material $ff_{def}$ in logarithmic coordinates for nanocomposites BSR/TC.

It is obvious, that the Eq. (35) is inapplicable to nanosystems, since $\psi \rightarrow 0$ assumes $K_T \rightarrow \infty$, that is physically incorrect. Therefore, the value $E_0$, obtained by the dependence $E(h_{pl})$ extrapolation (*see* Fig. 9) to $h_{pl}=0$, should be accepted as E for nanosystems [49].

Hence, the stated above results have shown, that elasticity modulus change at nanoindentation for particulate-filled elastomeric nanocomposites is due to a number of causes, which can be elucidated within the frameworks of anharmonicity conception and density fluctuation theory. Application of the first from the indicated conceptions assumes, that in nanocomposites during nanoindentation process local strain is realized, affecting polymer matrix only, and the transition to macrosystems means nanocomposite deformation as homogeneous system. The second from the mentioned conceptions has shown, that nano- and microsystems differ by density fluctuation absence in the first and availability of ones in the second. The last circumstance assumes, that for the considered nanocomposites density fluctuations take into account nanofiller and polymer matrix density difference. The transition from nano- to microsystems is realized in the case, when the deformed material volume exceeds nanofiller particles aggregate and surrounding it layers of polymer matrix combined volume [49].

In work [3], the following formula was offered for elastomeric nanocomposites reinforcement degree $E_n/E_m$ description:

$$\frac{E_n}{E_m} = 15.2\left[1-\left(d-d_{surf}\right)^{1/t}\right] \qquad (37)$$

where $t$ is index percolation, equal to 1.7 [28].

From the Eq. (37) it follows, that nanofiller particles (aggregates of particles) surface dimension $d_{surf}$ is the parameter, controlling nanocomposites reinforcement degree [53]. This postulate corresponds to the known principle about numerous division surfaces decisive role in nanomaterials as the basis of their properties change [54]. From the Eqs. (4)–(6) it follows unequivocally, that the value $d_{surf}$ is defined by nanofiller particles (aggregates of particles) size $R_p$ only. In its turn, from the Eq. (37) it follows, that elastomeric nanocomposites reinforcement degree $E_n/E_m$ is defined by the dimension $d_{surf}$ only, or, accounting for the said above, by the size $R_p$ only. This means, that the reinforcement effect is controlled by nanofiller particles (aggregates of particles) sizes only and in virtue of this is the true nanoeffect.

In Fig. 12, the dependence of $E_n/E_m$ on $(d-d_{surf})^{1/1.7}$ is adduced, corresponding to the Eq. (37), for nanocomposites with different elastomeric matrices (natural and butadiene-stirene rubbers, NR and BSR, accordingly) and different nanofillers (technical carbon of different marks, nano- and microshungite). Despite the indicated distinctions in composition, all adduced data are described well by the Eq. (37).

In Fig. 13, two theoretical dependences of $E_n/E_m$ on nanofiller particles size (diameter $D_p$), calculated according to the Eqs. (4)–(6) and (37), are adduced. However, at the curve 1 calculation the value $D_p$ for the initial nanofiller particles was used and at the curve 2 calculation – nanofiller particles aggregates size $D_p^{ag}$ (see Fig. 3). As it was expected [5], the growth $E_n/E_m$ at $D_p$ or $D_p^{ag}$ reduction, in addition the calculation with $D_p$ (nonaggregated nanofiller) using gives higher $E_n/E_m$ values in comparison with the aggregated one ($D_p^{ag}$ using). At $D_p\pounds50$ nm faster growth $E_n/E_m$ at $D_p$ reduction is observed than at $D_p>50$ nm, that was also expected. In Fig. 13, the critical theoretical value $D_p^{cr}$ for this transition, calculated according to the indicated above general principles [54], is pointed out by a vertical shaded line. In conformity with these principles the nanoparticles size in nanocomposite is determined according to the condition, when division surface fraction in the entire nanomaterial volume makes up about 50% and more. This fraction is estimated approximately by the ratio $3l_{if}/D_p$, where $l_{if}$ is interfacial layer thickness. As it was noted above, the data of Fig. 1 gave the average experimental value $l_{if}$»8.7 nm. Further from the condition $3l_{if}/D_p$»0.5 let us obtain $D_p$»52 nm that is shown in Fig. 13 by a vertical shaded line. As it was expected, the value $D_p$»52 nm is a boundary one for regions of slow ($D_p>52$ nm) and fast ($D_p\leq52$ nm) $E_n/E_m$ growth at $D_p$ reduction. In other words, the materials with nanofiller particles size $D_p\leq52$ nm ("superreinforcing" filler according to the terminology of paper [5]) should be considered true nanocomposites.

Let us note in conclusion, that although the curves 1 and 2 of Fig. 13 are similar ones, nanofiller particles aggregation, which the curve 2 accounts for, reduces essentially enough nanocomposites reinforcement degree. At the same time the experimental data correspond exactly to the curve 2, that was to be expected in virtue of aggregation processes, which always took place in real composites [4] (nanocomposites [55]). The values $d_{surf}$ obtained according to the Eqs. (4)–(6), correspond well to the determined experimentally ones. So, for nanoshungite and two marks of technical carbon the calculation by the indicated method gives the following

$d_{surf}$ values: 2.81, 2.78 and 2.73, whereas experimental values of this parameter are equal to: 2.81, 2.77 and 2.73, i.e., practically a full correspondence of theory and experiment was obtained.

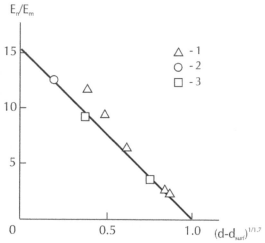

**FIGURE 12**   The dependence of reinforcement degree $E_n/E_m$ on parameter $(d-d_{surf})^{1/1.7}$ value for nanocomposites NR/TC (1), BSR/TC (2) and BSR/shungite (3).

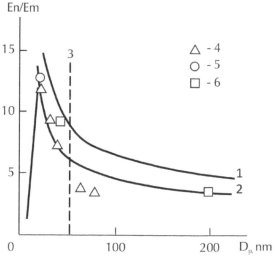

**FIGURE 13**   The theoretical dependences of reinforcement degree $E_n/E_m$ on nanofiller particles size $D_p$, calculated according to the Eqs. (4)–(6) and (37), at initial nanoparticles (1) and nanoparticles aggregates (2) size using. 3 – the boundary value $D_p$, corresponding to true nanocomposite. 4–6 – the experimental data for nanocomposites NR/TC (4), BSR/TC (5) and BSR/shungite (6).

## 2.4  CONCLUSIONS

Hence, the stated above results have shown, that the elastomeric reinforcement effect is the true nanoeffect, which is defined by the initial nanofiller particles size only. The indicated particles aggregation, always taking place in real materials, changes reinforcement degree quantitatively only, namely, reduces it. This effect theoretical treatment can be received within the frameworks of fractal analysis. For the considered nanocomposites the nanoparticle size upper limiting value makes up ~52 nm.

## KEYWORDS

- **Berkovich indentor**
- **nanoshungite**
- **Oliver-pharr method**
- **quadrates method**
- **superreinforcing**

## REFERENCES

1. Yanovskii Yu.G.; Kozlov, G. V.; Karnet Yu.N. Mekhanika Kompozitsionnykh Materialov i Konstruktsii, **2011**, *17(2)*, 203–208.
2. Malamatov A.Kh.; Kozlov, G. V.; Mikitaev, M. A. Reinforcement Mechanisms of Polymer Nanocomposites. Moscow, Publishers of the, D. I. Mendeleev RKhTU, **2006,** 240.
3. Mikitaev, A. K.; Kozlov, G. V.; Zaikov, G. E. Polymer Nanocomposites: Variety of Structural Forms and Applications. Moscow, Nauka, **2009,** 278.
4. Kozlov, G. V.; Yanovskii, Yu.G.; Karnet, Yu.N. Structure and Properties of Particulate-filled Polymer Composites: the Fractal Analysis. Moscow, Al'yanstrans-atom, **2008,** 363.
5. Edwards, J. D. C. *Mater. Sci.* **1990**, *25(12)*, 4175–4185.
6. Buchachenko, A. L. *Uspekhi Khimii*, **2003**, *72(5)*, 419–437.
7. Kozlov, G. V.; Yanovskii Yu.G.; Burya, A. I.; Aphashagova Z.Kh. Mekhanika Kompozitsionnykh Materialov i Konstruktsii, **2007**, *13(4)*, 479–492.
8. Lipatov, Yu.S. *The Physical Chemistry of Filled Polymers*. Moscow, Khimiya, **1977,** 304.

9. Bartenev, G. M.; Zelenev, Yu.V. Physics and Mechanics of Polymers. Moscow, Vysshaya Shkola, **1983**, 391.

10. Kozlov, G. V.; Mikitaev, A. K. Mekhanika Kompozitsionnykh Materialov i Konstruktsii, **1996**, *2(3–4)*, 144–157.

11. Kozlov, G. V.; Yanovskii Yu.G.; Zaikov, G. E. Structure and Properties of Particulate-filled Polymer Composites: The Fractal Analysis. New York, Nova Science Publishers, Inc.; **2010**, 282.

12. Mikitaev, A. K.; Kozlov, G. V.; Zaikov, G. E. Polymer Nanocomposites: Variety of Structural Forms and Applications. Nova Science Publishers, Inc.: New York, **2008**, 319.

13. McClintok, F. A.; Argon, A. S. Mechanical Behavior of Materials. Reading, Addison-Wesley Publishing Company, Inc.; **1966**, 440.

14. Kozlov, G. V.; Mikitaev, A. K. Doklady AN SSSR, **1987**, *294(5)*, 1129–1131.

15. Honeycombe, R. W. K. The Plastic Deformation of Metals. Boston, Edward Arnold (Publishers), Ltd.; **1968**, 398.

16. Dickie, R. A. In book: polymer Blends. 1. Academic Press: New York, San-Francisco, London, **1980**, 386–431.

17. Kornev, Yu.V.; Yumashev, O. B.; Zhogin A.; Karnet,Yu.N.; Yanovskii, Yu.G. Kautschuk i Rezina, **2008**, *6*, 18–23.

18. Oliver, W. C.; Pharr, J. G. M. *Mater. Res.* **1992**, *7(6)*, 1564–1583.

19. Kozlov, G. V.; Yanovskii, Yu.G.; Lipatov, Yu.S. Mekhanika Kompozitsionnykh Materialov i Konstruktsii, **2002**, *8(1)*, 111–149.

20. Kozlov, G. V.; Burya, A. I.; Lipatov, Yu.S. Mekhanika Kompozitnykh Materialov, **2006**, *42(6)*, 797–802.

21. Hentschel, H. G. E.; Deutch Phys, J. M.; Rev. A, **1984**, *29(3)*, 1609–1611.

22. Kozlov, G. V.; Ovcharenko, E. N.; Mikitaev, A. K. Structure of Polymers Amorphous State. Moscow, Publishers of the, D. I. Mendeleev RKhTU, **2009**, 392.

23. Yanovskii, Yu.G.; Kozlov, G. V.; Mater, VII Intern. Sci.-Pract. Conf. "New Polymer Composite Materials." *Nal'chik, KBSU*, **2011**, 189–194.

24. Wu, J S, Polymer Sci.: Part B: *Polymer Phys.;* **1989**, 27 *(4)*, 723–741.

25. Aharoni, S. M. *Macromolecules*, **1983**, *16 (9)*, 1722–1728.

26. Budtov The Physical Chemistry of Polymer Solutions. Sankt-Peterburg, Khimiya, **1992**, 384.

27. Aharoni Macromolecules, S. M.; **1985**, *18 (12)*, 2624–2630.

28. Bobryshev, A. N.; Kozomazov N.; Babin, L. O.; Solomatov I. Synergetics of Composite Materials. Lipetsk, NPO ORIUS, **1994**, 154.

29. Kozlov, G. V.; Yanovskii Yu.G.; Karnet Yu.N. Mekhanika Kompozitsionnykh Materialov i Konstruktsii, **2005**, *11 (3)*, 446–456.

30. Sheng N.; Boyce, M. C.; Parks, D. M.; Rutledge, G. C.; Abes, J. I.; Cohen, R. E.; *Polymer*, **2004**, *45 (2)*, 487–506.

31. Witten, T. A.; Meakin Phys P, *Rev. B*, **1983**, *28 (10)*, 5632–5642.

32. Witten, T. A.; Sander Phys, L. M.; *Rev. B*, **1983**, *27 (9)*, 5686–5697.

33. Happel J.; Brenner G. Hydrodynamics at Small Reynolds Numbers. Moscow, Mir, **1976**, 418.

34. Mills, J. N. *J. Appl. Polymer Sci.;* **1971**, *15(11)*, 2791–2805.

35. Kozlov, G. V.; Yanovskii Yu.G.; Mikitaev, A. K. Mekhanika Kompozitnykh Materialov, **1998,** *34(4),* 539–544.
36. Balankin, A. S. Synergetics of Deformable Body. Moscow, Publishers of Ministry Defence SSSR, **1991,** 404.
37. Hornbogen, E. *Intern. Mater. Res.;* **1989,** *34(6),* 277–296.
38. Pfeifer, P. *Appl. Surf. Sci.;* **1984,** *18(1),* 146–164.
39. Avnir, D.; Farin, D.; Pfeifer, J. P. *Colloid Interface Sci.;* **1985,** *103 (1),* 112–123.
40. Ishikawa, J. K. *Mater. Sci. Lett.;* **1990,** *9 (4),* 400–402.
41. Ivanova, S.; Balankin, A. S.; Bunin, I.Zh.; Oksogoev, A. A. Synergetics and Fractals in Material Science: Moscow, Nauka, **1994,** 383
42. ffstovskii, G. V.; Kolmakov, L. G.; Terent'ev V. E.; Metally, **1993,** *4,* 164–178.
43. Hansen J.; Skjeitorp Phys, A. T. *Rev. B,* **1988,** *38 (4),* 2635–2638.
44. Pfeifer; Avnir, D.; Farin, J. D. Stat. Phys.; **1984,** *36 (5/6),* 699–716.
45. Farin D.; Peleg S.; Yavin D.; Avnir Langmuir, D., **1985,** *1 (4),* 399–407.
46. Kozlov, G. V.; Sanditov, D. S. Anharmonical Effects and Physical-Mechanical Properties of Polymers: Novosibirsk, Nauka, **1994,** 261.
47. Bessonov, M. I.; Rudakov Vysokomolek, A. P.; Soed. B, **1971,** *13 (7),* 509–511.
48. Kubat, J.; Rigdahl, M.; Welander, J. M., *Appl. Polymer Sci.* **1990,** *39(5),* 1527–1539.
49. Yanovskii,Yu.G.; Kozlov, G. V.; Kornev, Yu.V.; Boiko, O. V.; Karnet Yu.N. Mekhanika Kompozitsionnykh Materialov i Konstruktsii, **2010,** *16 (3),* 445–453.
50. Yanovskii, Yu.G.; Kozlov, G. V.; Aloev Mater, V. Z.; Intern. Sci.-Pract. Conf. "Modern Problems of APK Innovation Development Theory and Practice." Nal'chik, KBSSKhA, **2011,** 434–437.
51. Chow Polymer, T. S.; **1991,** *32 (1),* 29–33.
52. Ahmed S.; Jones J. F. R., *Mater. Sci.* **1990,** *25 (12),* 4933–4942.
53. Kozlov, G. V.; Yanovskii Yu.G.; Aloev Mater, V. Z.; Intern. Sci.-pract. Conf.; dedicated to FMEP 50-th Anniversary. Nal'chik, KBSSKhA, **2011,** 83–89.
54. Andrievskii, R. A. Rossiiskii Khimicheskii Zhurnal, **2002,** *46 (5),* 50–56.
55. Kozlov, G. V.; Sultonov, N.Zh.; Shoranova, L. O.; Mikitaev, A. K. Naukoemkie Tekhnologii, **2011,** *12(3),* 17–22.

# CHAPTER 3

# SIMULATION OF NANOELEMENTS FORMATION AND INTERACTION

A. V. VAKHRUSHEV and A. M. LIPANOV

# CONTENTS

## 3.1   INTRODUCTION

The properties of a nanocomposite are determined by the structure and properties of the nanoelements, which form it. One of the main tasks in making nanocomposites is building the dependence of the structure and shape of the nanoelements forming the basis of the composite on their sizes. This is because with an increase or a decrease in the specific size of nanoelements (nanofibers, nanotubes, nanoparticles, etc.), their physical-mechanical properties such as coefficient of elasticity, strength, deformation parameter, etc. are varying over one order [1–5].

The calculations and experiments show that this is primarily due to a significant rearrangement (which is not necessarily monotonous) of the atomic structure and the shape of the nanoelement. The experimental investigation of the above parameters of the nanoelements is technically complicated and laborious because of their small sizes. In addition, the experimental results are often inconsistent. In particular, some authors have pointed to an increase in the distance between the atoms adjacent to the surface in contrast to the atoms inside the nanoelement, while others observe a decrease in the aforementioned distance [6].

Thus, further detailed systematic investigations of the problem with the use of theoretical methods, i.e., mathematical modeling, are required.

The atomic structure and the shape of nanoelements depend both on their sizes and on the methods of obtaining, which can be divided into two main groups:

1.  Obtaining nanoelements in the atomic coalescence process by "assembling" the atoms and by stopping the process when the nanoparticles grow to a desired size (the so-called "bottom-up" processes). The process of the particle growth is stopped by the change of physical or chemical conditions of the particle formation, by cutting off supplies of the substances that are necessary to form particles, or because of the limitations of the space where nanoelements form.

2.  Obtaining nanoelements by breaking or destruction of more massive (coarse) formations to the fragments of the desired size (the so-called "up down" processes).

In fact, there are many publications describing the modeling of the "bottom-up" processes [7, 8], while the "up down" processes have been studied very little. Therefore, the objective of this work is the investigation of the regularities of the changes in the structure and shape of nanoparticles formed in the destruction ("up down") processes depending on the nanoparticle sizes, and building up theoretical dependences describing the above parameters of nanoparticles.

When the characteristics of powder nanocomposites are calculated it is also very important to take into account the interaction of the nanoelements since the changes in their original shapes and sizes in the interaction process and during the formation (or usage) of the nanocomposite can lead to a significant change in its properties and a cardinal structural rearrangement. In addition, the experimental investigations show the appearance of the processes of ordering and self-assembling leading to a more organized form of a nanosystem [9–15]. In general, three main processes can be distinguished: the first process is due to the regular structure formation at the interaction of the nanostructural elements with the surface where they are situated; the second one arises from the interaction of the nanostructural elements with one another; the third process takes place because of the influence of the ambient medium surrounding the nanostructural elements. The ambient medium influence can have "isotropic distribution" in the space or it can be presented by the action of separate active molecules connecting nanoelements to one another in a certain order. The external action significantly changes the original shape of the structures formed by the nanoelements. For example, the application of the external tensile stress leads to the "stretch" of the nanoelement system in the direction of the maximal tensile stress action; the rise in temperature, vice versa, promotes a decrease in the spatial anisotropy of the nanostructures [10]. Note that in the self-organizing process, parallel with the linear moving, the nanoelements are in rotary movement. The latter can be explained by the action of moment of forces caused by the asymmetry of the interaction force fields of the nanoelements, by the presence of the "attraction" and "repulsion" local regions on the nanoelement surface, and by the "nonisotropic" action of the ambient as well.

The above phenomena play an important role in nanotechnological processes. They allow developing nanotechnologies for the formation of nanostructures by the self-assembling method (which is based on self-organizing processes) and building up complex spatial nanostructures consisting of different nanoelements (nanoparticles, nanotubes, fullerenes, super-molecules, etc.) [15]. However, in a number of cases, the tendency towards self-organization interferes with the formation of a desired nanostructure. Thus, the nanostructure arising from the self-organizing process is, as a rule, "rigid" and stable against external actions. For example, the "adhesion" of nanoparticles interferes with the use of separate nanoparticles in various nanotechnological processes, the uniform mixing of the nanoparticles from different materials and the formation of nanocomposite with desired properties. In connection with this, it is important to model the processes of static and dynamic interaction of the nanostructure elements. In this case, it is essential to take into consideration the interaction force moments of the nanostructure elements, which causes the mutual rotation of the nanoelements.

The investigation of the above dependences based on the mathematical modeling methods requires the solution of the aforementioned problem on the atomic level. This requires large computational aids and computational time, which makes the development of economical calculation methods urgent. The objective of this work was the development of such a technique.

This chapter gives results of the studies of problems of numeric modeling within the framework of molecular mechanics and dynamics for investigating the regularities of the amorphous phase formation and the nucleation and spread of the crystalline or hypocrystalline phases over the entire nanoparticle volume depending on the process parameters, nanoparticles sizes and thermodynamic conditions of the ambient. Also the method for calculating the interactions of nanostructural elements is offered, which is based on the potential built up with the help of the approximation of the numerical calculation results using the method of molecular dynamics of the pairwise static interaction of nanoparticles. Based on the potential of the pairwise interaction of the nanostructure elements, which takes into account forces and moments of forces, the method for calculating the ordering and self-organizing processes has been developed. The investigation results on the self-organization of the system consisting of two or more particles are presented and the analysis of the equilibrium stability of various types of nanostructures has been carried out. These results are a generalization of the authors' research in [16–24]. A more detailed description of the problem you can obtain in these works.

## 3.2   PROBLEM STATEMENT AND MODELING TECHNIQUE

The problem on calculating the internal structure and the equilibrium configuration (shape) of separate noninteracting nanoparticles by the molecular mechanics and dynamics methods has two main stages:

1.   The "initiation" of the task, i.e., the determination of the conditions under which the process of the nanoparticle shape and structure formation begins.
2.   The process of the nanoparticle formation.

Note that the original coordinates and initial velocities of the nanoparticle atoms should be determined from the calculation of the macroscopic parameters of the destructive processes at static and dynamic loadings taking place both on the nano-scale and on the macroscale. Therefore, in the general case, the coordinates and velocities are the result of solving the problem of modeling physical-mechanical destruction processes at different structural levels. This problem due to its enormity and complexity is not considered in this paper. The detailed description of its statement and the numerical results of its solution are given in the works of the authors [16–19].

The problem of calculating the interaction of ordering and self-organization of the nanostructure elements includes three main stages: the first stage is building the internal structure and the equilibrium configuration (shape) of each separate noninteracting nanostructure element; the second stage is calculating the pairwise interaction of two nanostructure elements; and the third stage is establishing the regularities of the spatial structure and evolution with time of the nanostructure as a whole.

Let us consider the above problems in sequence.

### 3.1.1   THE CALCULATION OF THE INTERNAL STRUCTURE AND THE SHAPE OF THE NONINTERACTING NANOELEMENT

The initialization of the problem is in giving the initial coordinates and velocities of the nanoparticle atoms

$$\vec{x}_i = \vec{x}_{i0}, \vec{V}_i = \vec{V}_{i0}, \ t = 0, \vec{x}_i \subset \Omega_k \qquad (1)$$

where $\vec{x}_{i0}, \vec{x}_i$ are original and current coordinates of the $i$-th atom; $\vec{V}_{i0}, \vec{V}_i$ are initial and current velocities of the $i$-th atom, respectively; $\Omega_k$ is an area occupied by the nanoelement.

The problem of calculating the structure and the equilibrium configuration of the nanoelement will be carried out with the use of the molecular dynamics method taking into consideration the interaction of all the atoms forming the nanoelement. Since, at the first stage of the solution, the nanoelement is not exposed to the action of external forces, it is taking the equilibrium configuration with time, which is further used for the next stage of calculations.

At the first stage, the movement of the atoms forming the nanoparticle is determined by the set of Langevin differential equations at the boundary conditions in Eq. (1) [25]

$$m_i \times \frac{d\vec{V}_i}{dt} = \sum_{j=1}^{N_k} \vec{F}_{ij} + \vec{F}_i(t) - \alpha_i m_i \vec{V}_i, \qquad i = 1, 2, .., N_k,$$

$$\frac{d\vec{x}_i}{dt} = \vec{V}_i, \qquad (2)$$

where $N_k$ is the number of atoms forming each nanoparticle; $m_i$ is the mass of the $i$-th atom; $\alpha_i$ is the "friction" coefficient in the atomic structure; $\vec{F}_i(t)$ is a random set of forces at a given temperature which is given by Gaussian distribution.

The interatomic interaction forces usually are potential and determined by the relation

$$\vec{F}_{ij} = -\sum_{1}^{n} \frac{\partial \Phi(\vec{\rho}_{ij})}{\partial \vec{\rho}_{ij}} \ , i = 1, 2, ..., N_k, \ j = 1, 2, ..., N_k, \qquad (3)$$

where $\vec{\rho}_{ij}$ is a radius-vector determining the position of the $i$-th atom relative to the $j$-th atom; $\Phi(\vec{\rho}_{ij})$ is a potential depending on the mutual positions of all the atoms; $n$ is the number of interatomic interaction types.

In the general case, the potential $\Phi(\vec{\rho}_{ij})$ is given in the form of the sum of several components corresponding to different interaction types:

$$\Phi(\vec{\rho}_{ij}) = \Phi_{cb} + \Phi_{va} + \Phi_{ta} + \Phi_{pg} + \Phi_{vv} + \Phi_{es} + \Phi_{hb} \qquad (4)$$

Here the following potentials are implied: $\Phi_{cb}$ – of chemical bonds; $\Phi_{va}$ – of valence angles; $\Phi_{ta}$ – of torsion angles; $\Phi_{pg}$ – of flat groups; $\Phi_{vv}$ – of Van der Waals contacts; $\Phi_{es}$ – of electrostatics; $\Phi_{hb}$ – of hydrogen bonds.

The above addends have different functional forms. The parameter values for the interaction potentials are determined based on the experiments (crystallography, spectral, calorimetric, etc.) and quantum calculations [25].

Giving original coordinates (and forces of atomic interactions) and velocities of all the atoms of each nanoparticle in accordance with Eq. (2), at the start time, we find the change of the coordinates and the velocities of each nanoparticle atoms with time from the equation of motion, Eq. (1). Since the nanoparticles are not exposed to the action of external forces, they take some atomic equilibrium configuration with time that we will use for the next calculation stage.

### 3.1.2   THE CALCULATION OF THE PAIRWISE INTERACTION OF THE TWO NANOSTRUCTURE ELEMENTS

At this stage of solving the problem, we consider two interacting nanoelements. First, let us consider the problem statement for symmetric nanoelements, and then for arbitrary shaped nanoelements.

First of all, let us consider two symmetric nanoelements situated at the distance $S$ from one another (Fig. 1 ) at the initial conditions

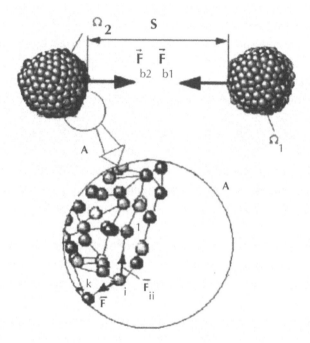

**FIGURE 1**   The scheme of the nanoparticle interaction; A – an enlarged view of the nanoparticle fragment.

$$\vec{x}_i = \vec{x}_{i0}, \vec{V}_i = 0, \; t = 0, \; \vec{x}_i \subset \Omega_1 \bigcup \Omega_2 \,, \tag{5}$$

where $\Omega_1, \Omega_2$ are the areas occupied by the first and the second nanoparticle, respectively.

We obtain the coordinates $\vec{x}_{i0}$ from Eq. (2) solution at initial conditions given in Eq. (1). It allows calculating the combined interaction forces of the nanoelements

$$\vec{F}_{b1} = -\vec{F}_{b2} = \sum_{i=1}^{N_1} \sum_{j=1}^{N_2} \vec{F}_{ij} \,, \tag{6}$$

where $i, j$ are the atoms and $N_1, N_2$ are the numbers of atoms in the first and in the second nanoparticle, respectively. Forces $\vec{F}_{ij}$ are defined from Eq. (3).

In the general case, the force magnitude of the nanoparticle interaction $\left|\vec{F}_{bi}\right|$ can be written as product of functions depending on the sizes of the nanoelements and the distance between them:

$$\left|\vec{F}_{bi}\right| = \Phi_{11}(S_{\tilde{n}}) \times \Phi_{12}(D) \tag{7}$$

The $\vec{F}_{bi}$ vector direction is determined by the direction cosines of a vector connecting the centers of the nanoelements.

Now, let us consider two interacting asymmetric nanoelements situated at the distance $S_c$ between their centers of mass (Fig. 2) and oriented at certain specified angles relative to each other.

In contrast to the previous problem, the interatomic interaction of the nanoelements leads not only to the relative displacement of the nanoelements but to their rotation as well. Consequently, in the general case, the sum of all the forces of the interatomic interactions of the nanoelements is brought to the principal vector of forces $\vec{F}_c$ and the principal moment $\vec{M}_c$

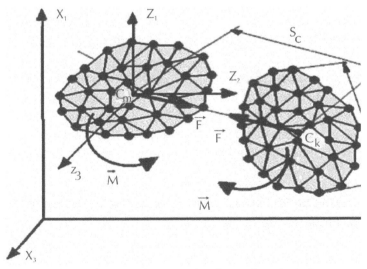

FIGURE 2  Two interacting nanoelements; $\vec{M}$, $\vec{F}$ are the principal moment and the principal vector of the forces, respectively.

$$\vec{F}_c = \vec{F}_{b1} = -\vec{F}_{b2} = \sum_{i=1}^{N_1} \sum_{j=1}^{N_2} \vec{F}_{ij} , \tag{8}$$

$$\vec{M}_c = \vec{M}_{c1} = -\vec{M}_{c2} = \sum_{i=1}^{N_1} \sum_{j=1}^{N_2} \vec{\rho}_{cj} \times \vec{F}_{ij} , \tag{9}$$

where $\vec{\rho}_{cj}$ is a vector connecting points $c$ and $j$.

The main objective of this calculation stage is building the dependences of the forces and moments of the nanostructure nanoelement interactions on the distance $S_c$ between the centers of mass of the nanostructure nanoelements, on the angles of mutual orientation of the nanoelements $\Theta_1, \Theta_2, \Theta_3$ (shapes of the nanoelements) and on the characteristic size D of the nanoelement. In the general case, these dependences can be given in the form:

$$\vec{F}_{bi} = \vec{\Phi}_F(S_c, \Theta_1, \Theta_2, \Theta_3, D) , \tag{10}$$

$$\vec{M}_{bi} = \vec{\Phi}_M(S_c, \Theta_1, \Theta_2, \Theta_3, D) , \tag{11}$$

For spherical nanoelements, the angles of the mutual orientation do not influence the force of their interaction; therefore, in Eq. (12), the moment is zero.

In the general case, Eqs. (11) and (12) can be approximated by analogy with (8), as the product of functions $S_0, \Theta_1, \Theta_2, \Theta_3, D$, respectively. For the further numerical solution of the problem of the self-organization of nanoelements, it is sufficient to give the above functions in their tabular form and to use the linear (or nonlinear) interpolation of them in space.

## 3.2 PROBLEM FORMULATION FOR INTERACTION OF SEVERAL NANOELEMENTS

When the evolution of the nanosystem as whole (including the processes of ordering and self-organization of the nanostructure nanoelements) is investigated, the movement of each system nanoelement is considered as the movement of a single whole. In this case, the trans-lational motion of the center of mass of each nanoelement is given in the coordinate system $X_1$, $X_2$ $X_3$, and the nanoelement rotation is described in the coordinate system $Z_1$, $Z_2$ $Z_3$, which is related to the center of mass of the nanoelement (Fig. 2). The system of equations describing the above processes has the form:

$$M_k \frac{d^2 X_1^k}{dt^2} = \sum_{j=1}^{N_e} F_{X_1}^{kj} + F_{X_1}^{ke},$$

$$M_k \frac{d^2 X_2^k}{dt^2} = \sum_{j=1}^{N_e} F_{X_2}^{kj} + F_{X_2}^{ke},$$

$$M_k \frac{d^2 X_3^k}{dt^2} = \sum_{j=1}^{N_e} F_{X_3}^{kj} + F_{X_3}^{ke}, \qquad (12)$$

$$J_{z_1}^k \frac{d^2 \Theta_1^k}{dt^2} + \frac{d\Theta_2^k}{dt} \times \frac{d\Theta_3^k}{dt} (J_{Z_3}^k - J_{Z_2}^k) = \sum_{j=1}^{N_e} M_{Z_1}^{kj} + M_{Z_1}^{ke},$$

$$J_{z_2}^k \frac{d^2 \Theta_2^k}{dt^2} + \frac{d\Theta_1^k}{dt} \times \frac{d\Theta_3^k}{dt} (J_{Z_1}^k - J_{Z_3}^k) = \sum_{j=1}^{N_e} M_{Z_2}^{kj} + M_{Z_2}^{ke},$$

$$J_{z_3}^k \frac{d^2 \Theta_3^k}{dt^2} + \frac{d\Theta_2^k}{dt} \times \frac{d\Theta_1^k}{dt} (J_{Z_2}^k - J_{Z_1}^k) = \sum_{j=1}^{N_e} M_{Z_3}^{kj} + M_{Z_3}^{ke},$$

where $X_i^k$, $\Theta_i^k$ are coordinates of the centers of mass and angles of the spatial orientation of the principal axis $Z_1$, $Z_2$ $Z_3$ of nanoelements; $F_{X_1}^{kj}, F_{X_2}^{kj}, F_{X_3}^{kj}$ are the interaction forces of nanoelements; $F_{X_1}^{ke}, F_{X_2}^{ke}, F_{X_3}^{ke}$ are external forces acting on nanoelements; $N_e$ is the number of nanoelements; $M_k$ is a mass of a nanoelement; $M_{Z_1}^{kj}, M_{Z_2}^{kj}, M_{Z_3}^{kj}$ is the moment of forces of the nanoelement interaction; $M_{Z_1}^{ke}, M_{Z_2}^{ke}, M_{Z_3}^{ke}$ are external moments acting on na-noelements; $J_{Z_1}^k, J_{Z_2}^k, J_{Z_3}^k$ are moments of inertia of a nanoelement.

The initial conditions for the system of Eqs. (13) and (14) have the form

$$\vec{X}^k = \vec{X}_0^k;\ \Theta^k = \Theta_0^k;\ \vec{V}^k = \vec{V}_0^k;\ \frac{d\Theta^k}{dt} = \frac{d\Theta_0^k}{dt};\ t = 0, \tag{13}$$

## 3.3.1 NUMERICAL PROCEDURES AND SIMULATION TECHNIQUES

In the general case, the problem formulated in the previous sections has no analytical solution at each stage; therefore, numerical methods for solving are used, as a rule. In this work, for the first stages, the numerical integration of the equation of motion of the nanoparticle atoms in the relaxation process are used in accordance with Verlet scheme [26]:

$$\vec{x}_i^{n+1} = \vec{x}_i^n + \Delta t\ \vec{V}_i^n + \left((\Delta t)^2/2m_i\right)\left(\sum_{j=1}^{N_k}\vec{F}_{ij} + \vec{F}_i - \alpha_i m_i \vec{V}_i^n\right)^n \tag{14}$$

$$\vec{V}_i^{n+1} = (1 - \Delta t\alpha_i)\vec{V}_i^n + (\Delta t/2m_i)((\sum_{j=1}^{N_k}\vec{F}_{ij} + \vec{F}_i)^n + (\sum_{j=1}^{N_k}\vec{F}_{ij} + \vec{F}_i)^{n+1}) \tag{15}$$

where $\vec{x}_i^n$, $\vec{V}_i^n$ are a coordinate and a velocity of the $i$-th atom at the $n$-th step with respect to the time; $\Delta t$ is a step with respect to the time.

The solution of the Eqs. (13) also requires the application of numerical methods of integration. In the present work, Runge–Kutta method [27] is used for solving Eq. (13).

$$(X_i^k)_{n+1} = (X_i^k)_n + (V_i^k)_n \Delta t + \frac{1}{6}(\mu_{1i}^k + \mu_{2i}^k + \mu_{3i}^k)\Delta t \tag{16}$$

$$(V_i^k)_{n+1} = (V_i^k)_n + \frac{1}{6}(\mu_{1i}^k + 2\mu_{2i}^k + 2\mu_{3i}^k + \mu_{4i}^k) \tag{17}$$

$$\mu_{1i}^k = \Phi_i^k(t_n;(X_i^k)_n,...;(V_i^k)_n...)\Delta t,$$

$$\mu_{2i}^k = \Phi_i^k(t_n + \frac{\Delta t}{2}; (X_i^k + V_i^k \frac{\Delta t}{2})_n,...;(V_i^k)_n + \frac{\mu_{1i}^k}{2},...)\Delta t \,,$$

$$\mu_{3i}^k = \Phi_i^k(t_n + \frac{\Delta t}{2}; (X_i^k + V_i^k \frac{\Delta t}{2} + \mu_{1i}^k \frac{\Delta t}{4})_n,...;(V_i^k)_n + \frac{\mu_{2i}^k}{2},...)\Delta t \qquad (18)$$

$$\mu_{4i}^k = \Phi_i^k(t_n + \Delta t; (X_i^k + V_i^k \Delta t + \mu_{2i}^k \frac{\Delta t}{2})_n,...;(V_i^k)_n + \mu_{2i}^k,...)\Delta t \,.$$

$$\Phi_i^k = \frac{1}{M_k}(\sum_{j=1}^{N_e} F_{X_3}^{kj} + F_{X_3}^{ke}) \qquad (19)$$

$$(\Theta_i^k)_{n+1} = (\Theta_i^k)_n + (\frac{d\Theta_i^k}{dt})_n \Delta t + \frac{1}{6}(\lambda_{1i}^k + \lambda_{2i}^k + \lambda_{3i}^k)\Delta t \qquad (20)$$

$$(\frac{d\Theta_i^k}{dt})_{n+1} = (\frac{d\Theta_i^k}{dt})_n + \frac{1}{6}(\lambda_{1i}^k + 2\lambda_{2i}^k + 2\lambda_{3i}^k + \lambda_{4i}^k) \qquad (21)$$

$$\lambda_{1i}^k = \Psi_i^k(t_n; (\Theta_i^k)_n,...;(\frac{d\Theta_i^k}{dt})_n...)\Delta t \,,$$

$$\lambda_{2i}^k = \Psi_i^k(t_n + \frac{\Delta t}{2}; (\Theta_i^k + \frac{d\Theta_i^k}{dt} \frac{\Delta t}{2})_n,...;(\frac{d\Theta_i^k}{dt})_n + \frac{\lambda_{1i}^k}{2},...)\Delta t \,,$$

$$\lambda_{3i}^k = \Psi_i^k(t_n + \frac{\Delta t}{2}; (\Theta_i^k + \frac{d\Theta_i^k}{dt} \frac{\Delta t}{2} + \lambda_{1i}^k \frac{\Delta t}{4})_n,...;(\frac{d\Theta_i^k}{dt})_n + \frac{\lambda_{2i}^k}{2},...)\Delta t \qquad (22)$$

$$\lambda_{4i}^k = \Psi_i^k(t_n + \Delta t; (\Theta_i^k + \frac{d\Theta_i^k}{dt} \Delta t + \lambda_{2i}^k \frac{\Delta t}{2})_n,...;(\frac{d\Theta_i^k}{dt})_n + \lambda_{2i}^k,...)\Delta t \,.$$

$$\Psi_1^k = \frac{1}{J_{Z_1}^k}(-\frac{d\Theta_2^k}{dt}\times\frac{d\Theta_3^k}{dt}(J_{Z1}^k - J_{Z_2}^k)+\sum_{j=1}^{N_e} M_{Z_1}^{kj} + M_{Z_1}^{ke}),$$

$$\Psi_2^k = \frac{1}{J_{Z_2}^k}(-\frac{d\Theta_1^k}{dt}\times\frac{d\Theta_3^k}{dt}(J_{Z_1}^k - J_{Z_3}^k)+\sum_{j=1}^{N_e} M_{Z_2}^{kj} + M_{Z_2}^{ke}) \qquad (23\text{-}a)$$

$$\Psi_3^k = \frac{1}{J_{Z_3}^k}(-\frac{d\Theta_1^k}{dt}\times\frac{d\Theta_2^k}{dt}(J_{Z_1}^k - J_{Z_2}^k)+\sum_{j=1}^{N_e} M_{Z_3}^{kj} + M_{Z_3}^{ke}),$$

were $i=1,2,3; \; k=1,2,\ldots N_e$

## 3.4   RESULTS AND DISCUSSIONS

Let us consider the realization of the above procedure taking as an example the calculation of the metal nanoparticle.

The potentials of the atomic interaction of Morse (Eq. (23b)) and Lennard-Johns (Eq. (24)) were used in the following calculations:

$$\Phi(\vec{\rho}_{ij})_m = D_m\,(\exp(-2\lambda_m(|\vec{\rho}_{ij}| - \rho_0)) - 2\exp(-\lambda_m(|\vec{\rho}_{ij}| - \rho_0))) \qquad (23b)$$

$$\Phi(\vec{\rho}_{ij})_{LD} = 4\varepsilon\left[\left(\frac{\sigma}{|\vec{\rho}_{ij}|}\right)^{12} - \left(\frac{\sigma}{|\vec{\rho}_{ij}|}\right)^6\right] \qquad (24)$$

where $D_m,\lambda_m,\rho_0,\varepsilon,\sigma$ are the constants of the materials studied.

For sequential and parallel solving the molecular dynamics equations, the program package developed at Applied Mechanics Institute, the Ural Branch of the Russian Academy of Sciences, and the advanced program package NAMD developed at the University of Illinois and Beckman Institute, USA, by the Theoretical Biophysics Group were used. The graphic imaging of the nanoparticle calculation results was carried out with the use of the program package VMD.

### 3.4.1 STRUCTURE AND FORMS OF NANOPARTICLES

At the first stage of the problem, the coordinates of the atoms positioned at the ordinary material lattice points (Fig. 3, Eq. (1)) were taken as the original coordinates. During the relaxation process, the initial atomic system is rearranged into a new "equilibrium" configuration (Fig. 3, Eq. (2)) in accordance with the calculations based on Eqs. (6)–(9), which satisfies the condition when the system potential energy is approaching the minimum (Fig. 3, the plot).

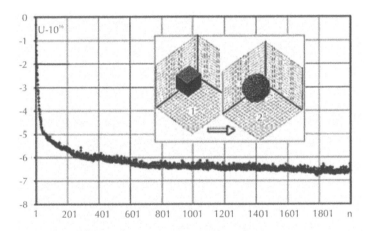

**FIGURE 3**   The initial crystalline (1) and cluster (2) structures of the nanoparticle consisting of 1331 atoms after relaxation; the plot of the potential energy $U$ [J] variations for this atomic system in the relaxation process ($n$ – number of iterations with respect to the time).

After the relaxation, the nanoparticles can have quite diverse shapes: globe-like, spherical centered, spherical eccentric, spherical icosahedral nanoparticles and asymmetric nanoparticles (Fig. 4).

In this case, the number of atoms N significantly determines the shape of a nanoparticle. Note, that symmetric nanoparticles are formed only at a certain number of atoms. As a rule, in the general case, the nanoparticle deviates from the symmetric shape in the form of irregular raised portions on the surface. Besides, there are several different equilibrium shapes for the same number of atoms. The plot of the nanoparticle potential energy change in the relaxation process (Fig. 5) illustrates it.

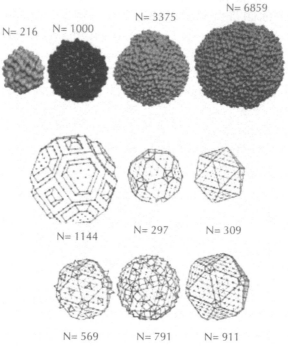

**FIGURE 4**   Nanoparticles of diverse shapes, depending on the number of atoms they contain.

As it follows from this figure, the curve has two areas: the area of the decrease of the potential energy and the area of its stabilization promoting the formation of the first nanoparticle equilibrium shape (1). Then, a repeated decrease in the nanoparticle potential energy and the stabilization area corresponding to the formation of the second nanoparticle equilibrium shape are observed (2). Between them, there is a region of the transition from the first shape to the second one (P). The second equilibrium shape is more stable due to the lesser nanoparticle potential energy. However, the first equilibrium shape also "exists" rather long in the calculation process. The change of the equilibrium shapes is especially characteristic of the nanoparticles with an "irregular" shape. The internal structure of the nanoparticles is of importance since their atomic structure significantly differs from the crystalline structure of the bulk materials: the distance between the atoms and the angles change, and the surface formations of different types appear. In Fig. 6, the change of the structure of a two-dimensional nanoparticle in the relaxation process is shown.

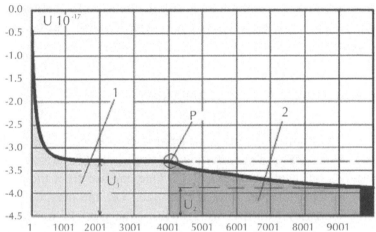

**FIGURE 5** The plot of the potential energy change of the nanoparticle in the relaxation process. 1 – a region of the stabilization of the first nanoparticle equilibrium shape; 2 – a region of the stabilization of the second nanoparticle equilibrium shape; P – a region of the transition of the first nanoparticle equilibrium shape into the second one.

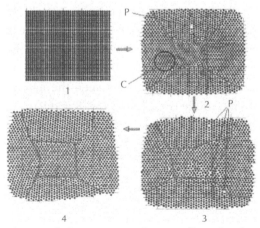

**FIGURE 6** The change of the structure of a two-dimensional nanoparticle in the relaxation process: 1 – the initial crystalline structure; 2, 3, 4 – the nanoparticles structures which change in the relaxation process; p – pores; c – the region of compression.

The figure shows how the initial nanoparticle crystalline structure (1) is successively rearranging with time in the relaxation process (positions 2, 3, 4). Note that the resultant shape of the nanoparticle is not round, i.e., it has "remembered" the initial atomic structure. It is also of interest that in the relaxation process, in the nanoparticle, the defects in the form of pores (designation "p" in the figure) and the density fluctuation regions (designation "c" in the figure) have been formed, which are absent in the final structure.

## 3.4.2   NANOPARTICLES INTERACTION

Let us consider some examples of nanoparticles interaction. Fig. 7 shows the calculation results demonstrating the influence of the sizes of the nanoparticles on their interaction force. One can see from the plot that the larger nanoparticles are attracted stronger, i.e., the maximal interaction force increases with the size growth of the particle. Let us divide the interaction force of the nanoparticles by its maximal value for each nanoparticle size, respectively. The obtained plot of the "relative" (dimensionless) force (Fig. 8) shows that the value does not practically depend on the nanoparticle size since all the curves come close and can be approximated to one line.

Figure 9 displays the dependence of the maximal attraction force between the nanoparticles on their diameter that is characterized by nonlinearity and a general tendency towards the growth of the maximal force with the nanoparticle size growth.

The total force of the interaction between the nanoparticles is determined by multiplying of the two plots (Figs. 8 and 9).

Using the polynomial approximation of the curve in Fig. 5 and the power mode approximation of the curve in Fig. 6, we obtain:

$$\overline{F} = (-1.13S^6 + 3.08S^5 - 3.41S^4 - 0.58S^3 + 0.82S - 0.00335)10^3 \tag{25}$$

$$F_{\max} \cdot = 0.5 \cdot 10^{-9} \cdot d^{1.499} \tag{26}$$

$$F = F_{max} \cdot \overline{F} \tag{27}$$

where $d$ and $s$ are the diameter of the nanoparticles and the distance between them [nm], respectively; $F_{max}$ is the maximal force of the interaction of the nanoparticles [N].

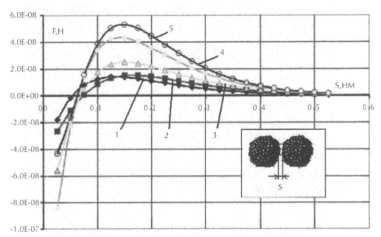

**FIGURE 7** The dependence of the interaction force F [N] of the nanoparticles on the distance S [nm] between them and on the nanoparticle size: $1 - d = 2.04$; $2 - d = 2.40$; $3 - d = 3.05$; $4 - d = 3.69$; $5 - d = 4.09$ [nm].

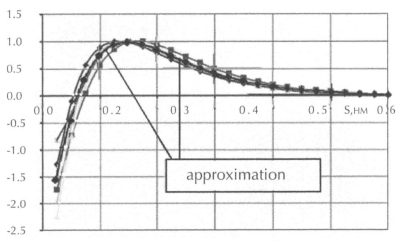

**FIGURE 8** The dependence of the "relative" force $\overline{F}$ of the interaction of the nanoparticles on the distance S [nm] between them.

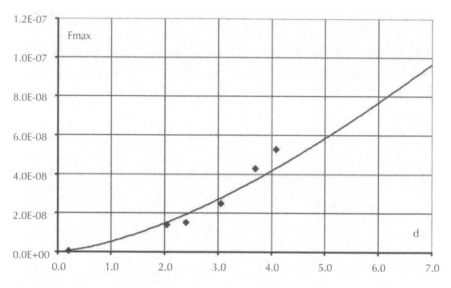

**FIGURE 9**   The dependence of the maximal attraction force $F_{max}$ [N] the nanoparticles on the nanoparticle diameter $d$ [nm].

Dependences of Eqs. (25) – (27) were used for the calculation of the nanocomposite ultimate strength for different patterns of nanoparticles' "packing" in the composite (Fig. 10).

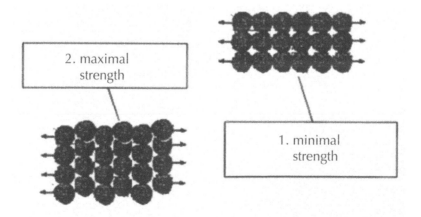

**FIGURE 10**   Different types of the nanoparticles' "packing" in the composite.

Figure 11 shows the dependence of the ultimate strength of the nano-composite formed by monodisperse nanoparticles on the nanoparticle sizes. One can see that with the decrease of the nanoparticle sizes, the ultimate strength of the nanomaterial increases, and vice versa. The calculations have shown that the nanocomposite strength properties are significantly influenced by the nanoparticles' "packing" type in the material. The material strength grows when the packing density of nanoparticles increases. It should be specially noted that the material strength changes in inverse proportion to the nanoparticle diameter in the degree of 0.5, which agrees with the experimentally established law of strength change of nanomaterials (the law by Hall-petch) [18]:

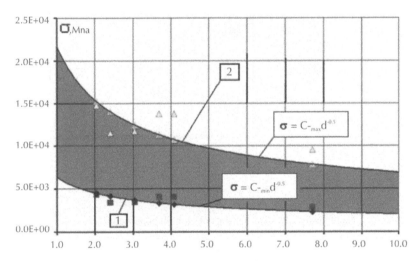

**FIGURE 11** The dependence of the ultimate strength $\sigma$ [MPa] of the nanocomposite formed by monodisperse nanoparticles on the nanoparticle sizes $d$ [nm].

$$\sigma = C \cdot d^{-0.5},\qquad(28)$$

where $C = C_{max} = 2.17 \times 10^4$ is for the maximal packing density; $C = C_{min} = 6.4 \times 10^3$ is for the minimal packing density.

The electrostatic forces can strongly change force of interaction of nanoparticles. For example, numerical simulation of charged sodium (NaCl) nanoparticles system (Fig. 12) has been carried out. Considered ensemble consists of eight separate nanoparticles. The nanoparticles interact due to Van-der-Waals and electrostatic forces.

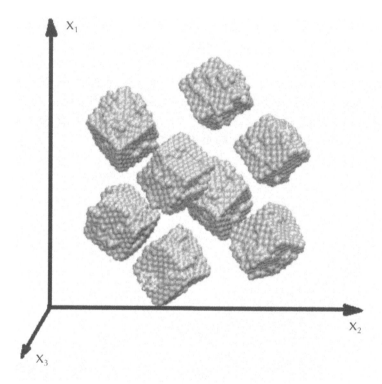

**FIGURE 12**   Nanoparticles system consists of eight nanoparticles NaCl.

Results of particles center of masses motion are introduced at Fig. 13 representing trajectories of all nanoparticles included into system. It shows the dependence of the modulus of displacement vector $|R|$ on time. One can see that nanoparticle moves intensively at first stage of calculation process. At the end of numerical calculation, all particles have got new stable locations, and the graphs of the radius vector $|R|$ become stationary. However, the nanoparticles continue to "vibrate" even at the final stage of numerical calculations. Nevertheless, despite of "vibration," the system of nanoparticles occupies steady position.

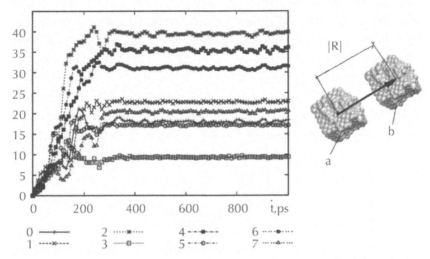

**FIGURE 13**   The dependence of nanoparticle centers of masses motion $|R|$ on the time t; a, b–the nanoparticle positions at time 0 and $t$, accordingly; 1–8 – are the numbers of the nanoparticles.

However, one can observe a number of other situations. Let us consider, for example, the self-organization calculation for the system consisting of 125 cubic nanoparticles, the atomic interaction of which is determined by Morse potential (Fig. 14).

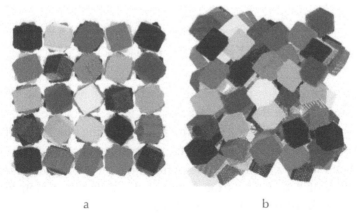

a                                                           b

**FIGURE 14**   The positions of the 125 cubic nanoparticles: (a) initial configuration; (b) final configuration of nanoparticles.

As you see, the nanoparticles are moving and rotating in the self-organization process forming the structure with minimal potential energy.

Let us consider, for example, the calculation of the self-organization of the system consisting of two cubic nanoparticles, the atomic interaction of which is determined by Morse potential [12]. Figure 15 displays possible mutual positions of these nanoparticles. The positions, where the principal moment of forces is zero, corresponds to pairs of the nanoparticles 2–3; 3–4; 2–5 (Fig. 15) and defines the possible positions of their equilibrium.

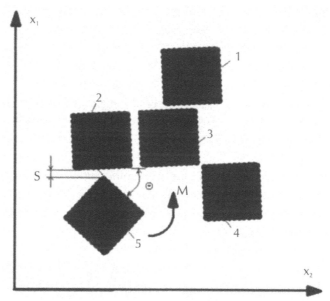

**FIGURE 15**   Characteristic positions of the cubic nanoparticles.

Figure 16 presents the dependence of the moment of the interaction force between the cubic nanoparticles 1–3 (Fig. 15) on the angle of their relative rotation. From the plot follows that when the rotation angle of particle 1 relative to particle 3 is $\pi/4$, the force moment of their interaction is zero. At an increase or a decrease in the angle the force moment appears. In the range of $\pi/8 < \theta < 3\pi/4$ the moment is small. The force moment rapidly grows outside of this range. The distance $s$ between the nanoparticles plays a significant role in establishing their equilibrium. If $s > s_0$ (where $s_0$ is the distance, where the interaction forces of the nanoparticles are zero), then the particles are attracted to one another. In this case, the sign of the moment corresponds to the sign of the angle $\theta$ deviation from $\pi/4$. At $S < S_0$ (the repulsion of the nanoparticles), the sign of the moment is opposite to the sign of the angle deviation. In other words, in the first case, the increase of the angle deviation causes the increase of the moment promoting the movement of the nanoelement in the given direction, and in the second case, the angle deviation causes the increase of the moment hindering the movement of the nanoelement in the given direction. Thus, the first case corresponds to the unstable equilibrium of nanoparticles, and the second case – to their stable equilibrium. The potential energy change plots for the system of the interaction of two cubic nanoparticles (Fig. 17) illustrate the influence of the parameter $S$. Here, curve 1 corresponds to the condition $S < S_0$ and it has a well-expressed minimum in the 0.3 < $\theta$ < 1.3 region. At $\theta$ < 0.3 and $\theta$ > 1.3, the interaction potential energy sharply increases, which leads to the return of the system into the initial equilibrium position. At $S > S_0$ (curves 2–5), the potential energy plot has a maximum at the $\theta = 0$ point, which corresponds to the unstable position.

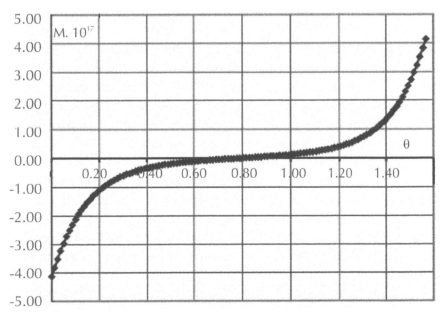

FIGURE 16   The dependence of the moment M [Nm] of the interaction force between cubic nanoparticles 1–3 (*see* Fig. 9) on the angle of their relative rotation $\theta$ [rad].

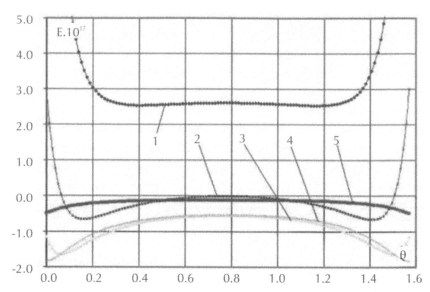

**FIGURE 17**   The plots of the change of the potential energy E [Nm] for the interaction of two cubic nanoparticles depending on the angle of their relative rotation $\theta$ [rad] and the distance between them (positions of the nanoparticles 1–3, Fig. 9).

The carried out theoretical analysis is confirmed by the works of the scientists from New Jersey University and California University in Berkeley who experimentally found the self-organization of the cubic microparticles of Plumbum Zirconate Titanate (PZT) [28]: the ordered groups of cubic microcrystals from PZT obtained by hydrothermal synthesis formed a flat layer of particles on the air-water interface, where the particle occupied the more stable position corresponding to position 2–3 in Fig. 15.

Thus, the analysis of the interaction of two cubic nanoparticles has shown that different variants of their final stationary state of equilibrium are possible, in which the principal vectors of forces and moments are zero. However, there are both stable and unstable stationary states of this system: nanoparticle positions 2–3 are stable, and positions 3–4 and 2–5 have limited stability or they are unstable depending on the distance between the nanoparticles.

Note that for the structures consisting of a large number of nanoparticles, there can be a quantity of stable stationary and unstable forms of equilibrium. Accordingly, the stable and unstable nanostructures of composite materials can appear. The search and analysis of the parameters determining the formation of stable nanosystems is an urgent task.

It is necessary to note, that the method offered has restrictions. This is explained by change of the nanoparticles form and accordingly variation of interaction pair potential during nanoparticles coming together at certain conditions.

The merge (accretion [4]) of two or several nanoparticles into a single whole is possible (Fig. 18). Change of a kind of connection cooperating nanoparticles (merging or coupling in larger particles) depending on its sizes, it is possible to explain on the basis of the analysis of the energy change graph of connection nanoparticles (Fig. 19). From Fig. 19 follows, that, though with the size increasing of a particle energy of nanoparticles connection $E_{np}$ grows also, its size in comparison with superficial energy $E_s$ of a particle sharply increases at reduction of the sizes nanoparticles. Hence, for finer particles energy of connection can appear sufficient for destruction of their configuration under action of a mutual attraction and merging in larger particle.

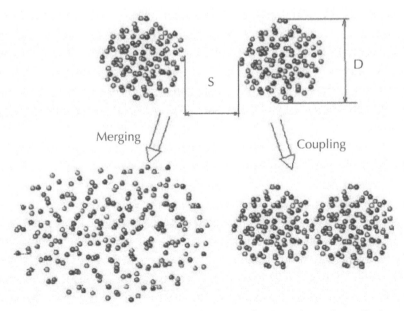

**FIGURE 18**  Different type of nanoparticles connection (merging and coupling).

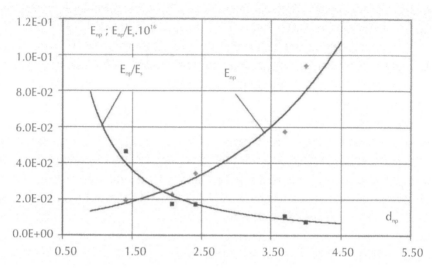

**FIGURE 19** Change of energy of nanoparticles connection $E_{np}$ [Nm] and $E_{np}$ ration to superficial energy $E_s$ depending on nanoparticles diameter $d$ [nm]. Points designate the calculated values. Continuous lines are approximations.

Spatial distribution of particles influences on rate of the forces holding nanostructures, formed from several nanoparticles, also. On Fig. 20 the chain nanoparticles, formed is resulted at coupling of three nanoparticles (from 512 atoms everyone), located in the initial moment on one line. Calculations have shown, that in this case nanoparticles form a stable chain. Thus, particles practically do not change the form and cooperate on "small platforms.".

In the same figure the result of connection of three nanoparticles, located in the initial moment on a circle and consisting of 256 atoms everyone is submitted. In this case particles incorporate among themselves "densely," contacting on a significant part of the external surface.

Distance between particles at which they are in balance it is much less for the particles collected in group ($L^0_{3np} < L^0_{2np}$) It confirms also the graph of forces from which it is visible, that the maximal force of an attraction between particles in this case (is designated by a continuous line) in some times more, than at an arrangement of particles in a chain (dashed line) $F_{3np} > F_{2np}$.

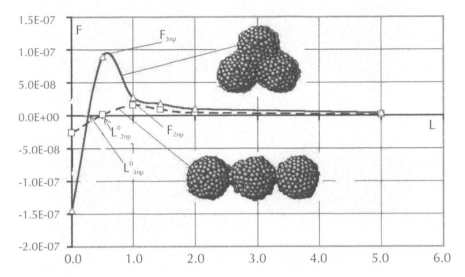

**FIGURE 20**   Change of force $F$ [N] of 3 nanoparticles interaction, consisting of 512 atoms everyone, and connected among themselves on a line and on the beams missing under a corner of 120 degrees, accordingly, depending on distance between them $L$ [nm].

Experimental investigation of the spatial structures formed by nanoparticles [4], confirm that nanoparticles gather to compact objects. Thus the internal nuclear structure of the connections area of nanoparticles considerably differs from structure of a free nanoparticle.

Nanoelements kind of interaction depends strongly on the temperature. In Fig. 21 shows the picture of the interaction of nanoparticles at different temperatures (Fig. 22). It is seen that with increasing temperature the interaction of changes in sequence: coupling (1,2), merging (3,4). With further increase in temperature the nanoparticles dispersed.

In the conclusion to the section we will consider problems of dynamics of nanoparticles. The analysis of interaction of nanoparticles among themselves also allows to draw a conclusion on an essential role in this process of energy of initial movement of particles. Various processes at interaction of the nanoparticles, moving with different speed, are observed: the processes of agglomerate formation, formation of larger particles at merge of the smaller size particles, absorption by large particles of the smaller ones, dispersion of particles on separate smaller ones or atoms.

**FIGURE 21**   Change of nanoparticles connection at increase in temperature.

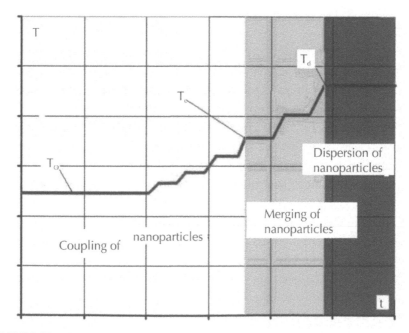

**FIGURE 22**   Curve of temperature change.

For example, in Fig. 23 the interactions of two particles are moving towards each other with different speed are shown. At small speed of moving is formed steady agglomerate (Fig. 23).

In the figure, Fig. 23 (left) is submitted interaction of two particles moving towards each other with the large speed. It is visible, that steady formation in this case is not appearing and the particles collapse.

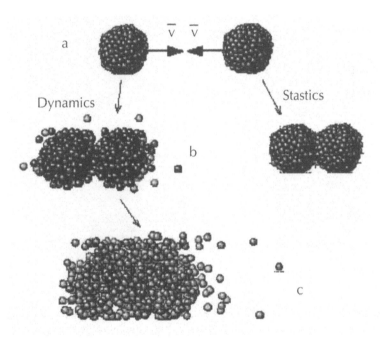

**FIGURE 23** Pictures of dynamic interaction of two nanoparticles: a) an initial configuration nanoparticles; b) nanoparticles at dynamic interaction, c) the "cloud" of atoms formed because of dynamic destruction two nanoparticles.

Nature of interaction of nanoparticles, along with speed of their movement, essentially depends on a ratio of their sizes. In Fig. 24 pictures of interaction of two nanoparticles of zinc of the different size are presented. As well as in the previous case of a nanoparticle move from initial situation (1) towards each other. At small initial speed, nanoparticles incorporate at contact and form a steady conglomerate (2).

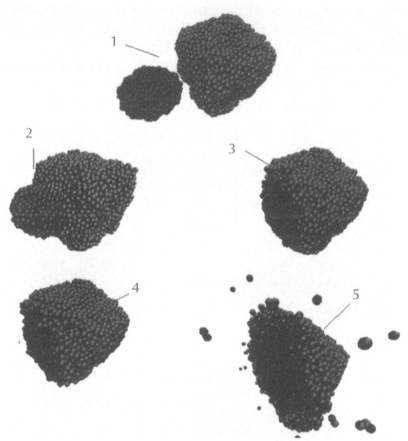

**FIGURE 24**    Pictures of interaction of two nanoparticles of zinc: 1 – initial configuration of nanoparticles; 2 – connection of nanoparticles, 3,4 – absorption by a large nanoparticle of a particle of the smaller size, 5 – destruction of nanoparticles at blow.

At increase in speed of movement larger nanoparticle absorbs smaller, and the uniform nanoparticle (3,4) is formed. At further increase in speed of movement of nanoparticles, owing to blow, the smaller particle intensively takes root in big and destroys its.

The given examples show that use of dynamic processes of pressing for formation of nanocomposites demands a right choice of a mode of the loading providing integrity of nanoparticles. At big energy of dynamic loading, instead of a nanocomposite with dispersion corresponding to the initial size of nanoparticles, the nanocomposite with much larger grain that will essentially change properties of a composite can be obtained.

## 3.5  CONCLUSIONS

In conclusion, the following basic regularities of the nanoparticle formation and self-organization should be noted.

1. The existence of several types of the forms and structures of nanoparticles is possible depending on the thermodynamic conditions.
2. The absence of the crystal nucleus in small nanoparticles.
3. The formation of a single (ideal) crystal nucleus defects on the nucleus surface connected to the amorphous shell.
4. The formation of the polycrystal nucleus with defects distributed among the crystal grains with low atomic density and change interatomic distances. In addition, the grain boundaries are nonequilibrium and contain a great number of grain-boundary defects.
5. When there is an increase in sizes, the structure of nanoparticles is changing from amorphous to roentgen-amorphous and then into the crystalline structure.
6. The formation of the defect structures of different types on the boundaries and the surface of a nanoparticle.
7. The nanoparticle transition from the globe-shaped to the crystal-like shape.
8. The formation of the "regular" and "irregular" shapes of nanoparticles depending on the number of atoms forming the nanoparticle and the relaxation conditions (the rate of cooling, first of all).
9. The structure of a nanoparticle is strained because of the different distances between the atoms inside the nanoparticle and in its surface layers.
10. The systems of nanoparticles can to form stable and unstable nanostructures.

## ACKNOWLEDGMENTS

This work was carried out with financial support from the Research Program of the Ural Branch of the Russian Academy of Sciences: the projects 12-P-12-2010, 12-C-1-1004, 12-T-1-1009 and was supported by the grants of Russian Foundation for Basic Research (RFFI) 04-01-96017-r2004ural_a; 05-08-50090-a; 07-01-96015-r_ural_a; 08-08-12082-ofi; 11-03-00571-a.

The author is grateful to his young colleagues Dr. A.A. Vakhrushev and Dr. A.Yu. Fedotov for active participation in the development of the software complex, calculations and analyzing of numerous calculation results.

The calculations were performed at the Joint Supercomputer Center of the Russian Academy of Sciences.

## KEYWORDS

- **Assembling**
- **Isotropic distribution**
- **Monodisperse nanoparticles**
- **Small platforms**

## REFERENCES

1. Qing-Qing Ni; Yaqin Fu; Masaharu Iwamoto, *"Evaluation of Elastic Modulus of Nano Particles in PMMA/Silica Nanocomposites"*, J. the Society of Materials Science, Japan, **2004,** *53(9),* 956–961,
2. Ruoff, R. S.; Nicola Pugno, M., *"Strength of nanostructures"*, Mechanics of the twenty-first century. Proceeding **of the 21-th international congress of theoretical and applied mechanics.** Warsaw: **Springer, 2004,** 303–311.
3. **Diao, J.; Gall, K.; Dunn M. L,** *"Atomistic simulation of the structure and elastic properties of gold nanowires"*, J. the Mechanics and Physics of Solids, **2004,** *52(9),* 1935–1962.
4. **Dingreville, R, Qu J, Cherkaoui M,** *"Surface free energy and its effect on the elastic behavior of nano-sized particles, wires and films"*, J. the Mechanics and Physics of Solids, **2004,** *53(8),* 1827–1854.

5. **Duan, H. L.; Wang J, Huang, Z. P.; Karihaloo, B. L.** "*Size-dependent effective elastic constants of solids containing nano-inhomogeneities with interface stress*", J. the Mechanics and Physics of Solids, **2005**, *53(7)*, 1574–1596.

6. Gusev, I. A.; Rempel, A. A. "*Nanocrystalline materials*" Moscow: physical mathematical literature, **2001**, (in Russian).

7. Hoare, M. R. "Structure and dynamics of simple microclusters," *Ach. Chem. Phys.;* **1987**, *40*, 49–135.

8. *Brooks, B. R.; Bruccoleri, R. E.; Olafson, B. D.; States, D. J.; Swaminathan, S.; Karplus, M.* "*CHARMM: a program for macromolecular energy minimization, and dynamics calculations*", Comput J. Chemistry, **1983**, *4(2)*, 187–217.

9. Friedlander, S. K. "*Polymer-like behavior of inorganic nanoparticle chain aggregates*", J. Nanoparticle Res., **1999**, *1*, 9–15.

10. Grzegorczyk, M.; Rybaczuk, M.; Maruszewski, K. "*Ballistic aggregation: an alternative approach to modeling of silica sol–gel structures*" Chaos, Solitons and Fractals, **2004**, *19*, 1003–1011.

11. Shevchenko E. V.; Talapin, D. V.; Kotov, N. A.; O'Brien, S.; Murray, C. B. "Structural diversity in binary nanoparticle superlattices," *Nature Lett.*, **2006**, *439*, 55–59.

12. Kang, Z. C.; Wang, Z. L. "On Accretion of nanosize carbon spheres", *J. Phys. Chemi.*, **1996**, *100*, 5163–5165.

13. Melikhov, I.; Bozhevol'nov, E. V.; "*Variability and self-organization in nanosystems*", J. Nanoparticle Res., **2003**, *5*, 465–472,

14. Kim, D.; Lu, W. "*Self-organized nanostructures in multiphase epilayers*" Nanotechnology, **2004**, *15*, 667–674.

15. Kurt Geckeler, E. "*Novel supermolecular nanomaterials: from design to reality*", Proceeding of the 12 Annual International Conference on Composites/Nano Engineering, Tenerife, Spain, August 1–6, CD Rom Edition, 2005.

16. Vakhrouchev, A.; Lipanov, A. M. "*A numerical analysis of the rupture of powder materials under the power impact influence*", Computer and Structures, **1992**, *1/2 (44)*, 481–486.

17. Vakhrouchev, A. V. "*Modeling of static and dynamic processes of nanoparticles interaction*", CD-ROM Proceeding **of the 21-th international congress of theoretical and applied mechanics,** ID12054, Warsaw, Poland, 2004.

18. Vakhrouchev, A. V.; "*Simulation of nanoparticles interaction*", Proceeding of the 12 Annual International Conference on Composites/Nano Engineering, Tenerife, Spain, August 1–6, CD Rom Edition, 2005.

19. Vakhrouchev, A. "Simulation of nano-elements interactions and self-assembling", Modeling and Simulation in Materials Science and Engineering, **2006**, *14*, 975–991.

20. Vakhrouchev A.; Lipanov, M. A. Numerical Analysis of the Atomic structure and Shape of Metal Nanoparticles,Computational Mathematics and Mathematical Physics, **2007**, *47 (10)*, 1702–1711.

21. Vakhrouchev A. Modelling of the Process of Formation and Use of Powder Nanocomposites, Composites with Micro and Nano-Structures. Computational Modeling and Experiments. Computational Methods in Applied Sciences Series. – Barcelona, Spain: Springer Science. **2008**, *9*, 107–136.

22. Vakhrouchev A. Modeling of the nanosystems formation by the molecular dynamics, mesodynamics and continuum mechanics methods, Multidiscipline Modeling in Material and Structures. **2009,** *5(2)*, 99–118.

23. Vakhrouchev A. Theoretical bases of nanotechnology application to thermal engines and equipment. – Izhevsk, Institute of Applied Mechanics, Ural Branch of the Russian Academy of Sciences, **2008,** 212p. (In Russian)

24. Alikin N.; Vakhrouchev A. V.; Golubchikov B.; Lipanov, A. M.; Serebrennikov, S. Y. Development and investigation of the aerosol nanotechnology. Moscow: mashinostroenie, 2010, 196 p.; (in Russian)

25. Heerman, W. D. *"Computer simulation methods in theoretical physics"* Berlin: Springer-Verlag, 1986.

26. Verlet, L., *"Computer "experiments" on classical fluids I. Thermo dynamical properties of Lennard-Jones molecules",* Phys. Rev.; **1967,** *159,* 98–103.

27. Korn, A. G.; Korn, M. T., *"Mathematical handbook",* New York: Mcgraw-Hill Book Company, **1968.**

28. Self-organizing of microparticles piezoelectric materials, News of chemistry, date-news.php.htm.

# CHAPTER 4

# ADVANCED POLYMERS

# SECTION I: A NOTE ON MACROMOLECULAR COIL CONNECTIVITY DEGREE

ALBERT H. C. WONG, IRVING I. GOTTESMAN, and ARTURAS PETRONIS

## CONTENTS

**SECTION III**

## 4.1  INTRODUCTION

As it is known, autohesion strength (coupling of the identical material surfaces) depends on interactions between some groups of polymers and treats usually in purely chemical terms on a qualitative level [1, 2]. In addition, the structure of neither polymer in volume nor its elements (for the example, macromolecular coil) is taken into consideration. The authors [3] showed that shear strength of autohesive joint $\tau_s$ depended on macromolecular coils contacts number $N_c$ on the boundary of division polymer-polymer. This means, that value $\tau_s$ is defined by the macromolecular coil structure, which can be described within the frameworks of fractal analysis with the help of three dimensions: fractal (Hausdorff) $D_f$, spectral (fraction) $d_s$ and the dimension of Euclidean space $d$, in which fractal is considered [4]. As it is known [5], the dimension $d_s$ characterizes macromolecular coil connectivity degree and varies from 1.0 for linear chain up to 1.33 for very branched macromolecules. In connection with this the question arises, how the value $d_s$ influences on autohesive joint strength $\tau_s$ or, in other words, what polymers are more preferable for the indicated joint formation – linear or branched ones. The purpose of the present communication is theoretical investigation of this effect within the frameworks of fractal analysis.

## 4.2  THEORETICAL ANALYSIS

The authors [3] showed that the value $\tau_s$ depended on $N_c$ as follows:

$$\ln \tau_s = 0.10 N_c - 6.0 \tag{1}$$

In its turn, macromolecular coils contacts number $N_c$ on boundary of division polymer-polymer can be determined according to the relationship [6]:

$$N_c \sim R_g^{D_{f1}+D_{f2}-d} \tag{2}$$

where $D_{f_1}$ and $D_{f_2}$ are dimensions of macromolecular coils of contacting polymers. In the autohesion case $D_{f_1}=D_{f_2}=D_f$ and then the Eq. (2) is reduced to the form [3]:

$$N_c \sim R_g^{2D_f-d} \tag{3}$$

The indicated above dimensions $D_f$, $d_s$ and $d$, characterizing macromolecular coil structure, are interconnected between themselves by the Eq. (4) [6]:

$$D_f = \frac{d_s(d+2)}{d_s+2} \tag{4}$$

The Eqs. (3) and (4) combination allows to obtain the direct dependence of $N_c$ and, hence, $\tau_s$ on dimension $d_s$:

$$N_c \sim R_g^{\frac{d_s d+4d_s-2d}{d_s+2}} \tag{5}$$

that in the most widespread case of three-dimensional Euclidean space $(d=3)$ gives still simpler relationship:

$$N_c \sim R_g^{(7d_s-6)/(d_s+2)} \tag{6}$$

The gyration radius value $R_g$ of macromolecular coil is connected with its fractal dimension $D_f$ by the relationship [6]:

$$R_g \sim N_{pol}^{1/D_f} \tag{7}$$

where $N_{pol}$ is polymerization degree, characterizing polymer molecular weight.

The Eqs. (4) and (7) combination at $d=3$ gives the dependence of $R_g$ on $d_s$ only:

$$R_g \sim N_{pol}^{(d_s+2)/5d_s} \tag{8}$$

At the value $R_g$ choice two variants are possible. The first from them assumes $N_{pol}$=2000, calculation $R_g$ according to the Eq. (8) at $d_s$=1.33 and further the condition $R_g$=const usage. The second one uses the Eq. (8) at $d_s$ variation within the range of 1.0–1.33 and $N_{pol}$=2000.

## 4.3   RESULTS AND DISCUSSION

In Fig. 1 the dependences of $\tau_s$ on $d_s$ at two conditions are adduced: $r_g$=const=44.7 relative units and $R_g$=variant (calculation according to the Eq. (8) at $d_s$=1.0–1.33). The plots of Fig. 1 allow to make two conclusions. Firstly, the estimation $R_g$ variant does not influence essentially on autohesive joint strength. Secondly, what most important is, the dependence of $\tau_s$ on $d_s$ within the range of $d_s$=1.0–1.20 (weakly branched polymer chains) is weak enough and absolute values $\tau_s$ are small and at $d_s$>1.20 sharp growth $\tau_s$ begins, which in the range of $d_s$=1.20–1.33 increases more than on one order. Hence, the data of Fig. 1 are supposed, that strongly branched polymer chains are ensured much higher autohesion level, than weakly branched ones.

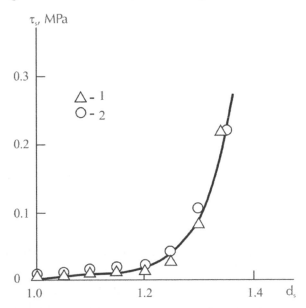

**FIGURE 1**   The dependences of shear strength of autohesive joint $\tau_s$ on spectral dimension $d_s$ at $R_g$=const (1) and $R_g$ calculation according to the Eq. (8) (2).

In Fig. 2 the dependences of $\tau_s$ on $d_s$, calculated according to the Eqs. (1), (5) and (8), are adduced for two $N_{pol}$ values: 2,000 and 4,000. As a matter of fact, this means two-fold increase of polymer molecular weight in the last case. The small distinction of $\tau_s$ for weakly branched polymer chains ($d_s \leq 1.20$) is again observed, but for strongly branched macromolecules ($d_s = 1.33$) molecular weight two-fold increase results to $\tau_s$ enhancement in 6.3 times.

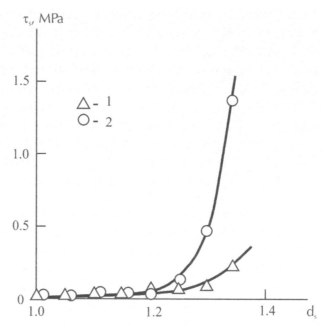

FIGURE 2    The dependence of shear strength of autohesive joint $\tau_s$ on spectral dimension $d_s$ at $N_{pol} = 2,000$ (1) and 4,000 (2).

## 4.4   CONCLUSIONS

Hence, the fulfilled in the present communication theoretical estimations have demonstrated, that small branching degree of chain ($d_s = 1.0$–$1.20$) practically does not influence on autohesion level and gives small absolute values of autohesive joint strength. However, strongly branched macromolecules ($d_s > 1.20$) usage can result to the indicated strength increase more than on one order. The molecular weight essential increase (in two times) can give more than sixfold growth of autohesive strength in case of very branched polymers. Let us note, that for the real polymers the chain elements, possessing strong specific interactions (e.g., hydrogen-bonded ones) [1] or bulk side groups, can play side branches role.

## KEYWORDS

- autohesion strength
- autohesive joint
- branched polymers
- Euclidean space

## REFERENCES

1. Boiko, Yu.M. Mekhanika Kompozitnykh Materialov, **2000**, *36(1)*, 127–134.
2. Voyutskii, S. S. Autohesion and Adhesion of High Polymers. Moscow, Postechizdat, **1960**, 244.
3. Mikitaev, A. K.; Kozlov, G. V.; Zaikov, G. E. Polymer Nanocomposites: variety of Structural Forms and Applications. New York, Nova Science Publishers, Inc.; **2008**, 319.
4. Rammal, R.; Toulouse, J. G.; *Phys. Lett. (Paris)*, **1983**, *44(1)*, L13–L22.
5. Alexander S.; Orbach J. R., *Phys. Lett. (Paris)*, **1982**, *43(7)*, L625–L631.
6. Vilgis, T. A. *Physica A*, **1988**, *153(2)*, 341–354.

# SECTION II: ON NONLINEAR DYNAMIC PROCESSES IN CNTs

S.A. SUDORGIN, M.B. BELONENKO, and N.G. LEBEDEV

## 4.1  INTRODUCTION

The rapid development of laser and optical technologies, the application of powerful lasers and the unique precision of optical measurements allowed substantial progress in the study of nonlinear phenomena that occur in a wide range of substances_with a very different physical properties [1,2]. This is due both to the rapid progress of computer technology, and with interest in modern physics the study of nonlinear dynamic processes. The object of the study should be as a substance with pronounced nonlinear properties and important from the viewpoint of practical applications. In the last decade, as a matter of researchers are increasingly attracted to carbon nanotubes (CNTs) are unique macromolecular systems [3]. Nanometer diameter and micron length of CNTs make them attractive for use in nano- and microelectronics, as they allow them to be considered the closest in structure to the ideal one-dimensional systems. The study of optical soliton propagation in CNTs is one of the most promising areas of research [4–9]. Although these papers were predicted by the existence possibility of electromagnetic solitons and dependence of their characteristics on the parameters of CNTs remained a series of questions requiring clarification. This question related to output beyond the one-dimensional approximation, and the question of the propagation of optical pulses with light diffraction, and the scattering problem of the optical pulse in the multidimensional case the inhomogeneities of various types. These structures are localized in three spatial dimensions, will be called "light bullets," because in this problem localization occurs in three dimensions. All these factors make the problem of the study of nonlinear dynamic processes in CNTs relevant for theory and in practice.

## 4.2   MODEL AND BASIC RELATIONS

We assume in the model that the electric field vector $\mathbf{E}(x, y, z, t)$ directed along the tube axis $x$, and the electromagnetic wave is moving in the transverse direction (Fig. 1). Carbon nanotubes are considered to be ideal to have a structural modification of the "zigzag" and located on the same interatomic distances equal to 0.34 nm [3] to simplify the calculations. Method of packing CNTs in the array is not important, because it does not take into account the interaction between the CNTs, and Therefore, used a tetragonal packing of single-walled carbon nanotubes in two-dimensional array with space group P42/mc($D_{4h}^9$) in Fig. 1 [3].

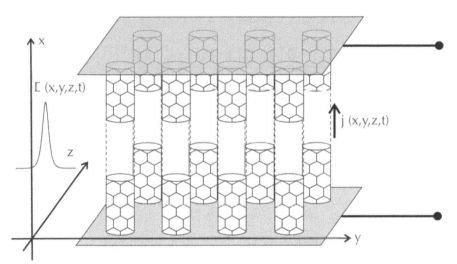

**FIGURE 1**   The geometry of the problem. Ultrashort optical pulse with the electric field $\mathbf{E}(x, y, z, t)$ is directed along the tube axis $x$, moving in the transverse direction along the axis $z$.

In this paper we ignore the influence of the substrate on which the nanotubes are grown. This can lead particularly to the appearance of surface waves of these systems, we will not considered these features in such article.

Electronic structure of "zigzag" carbon nanotubes is characterized by chiral indices $(n, 0)$ and described by the well-known dispersion relation which obtained in the framework of Huckel $\pi$-electron approximation [3]:

$$E(p) = \pm\gamma\sqrt{1 + 4\cos(ap_x)\cos(\pi s / n) + 4\cos^2(\pi s / n)} \qquad (11)$$

where $a = 3d / 2\hbar$, $d = 0.142$ nm is the distance between adjacent carbon atoms in graphene, $\mathbf{p} = (p_x, s)$ is the quasimomentum of the electrons in graphene, $p_x$ is the parallel component of the graphene sheet of the quasimomentum and $s = 1, 2, \ldots, n$ are the quantization numbers of the momentum components depending on the width of the graphene ribbon. Different signs are related to the conductivity band and to the valence band accordingly.

The electromagnetic field pulse is described classically on the basis of Maxwell's equations. We are considering the dielectric and magnetic properties of CNTs [10] and two-dimensionality of the problem the Maxwell equations for the vector potential $\mathbf{A}$ in the gage $E = -\dfrac{1}{c}\dfrac{\partial A}{\partial t}$ will be:

$$\frac{\partial^2 A}{\partial x^2} + \frac{\partial^2 A}{\partial y^2} + \frac{\partial^2 A}{\partial z^2} - \frac{1}{c}\frac{\partial^2 A}{\partial t^2} = -\frac{4\pi}{c}j \qquad (2)$$

We ignore the diffraction spreading of the laser beam in the $x$-direction in Eq. (2). The vector potential $\mathbf{A}$ is chosen as $\mathbf{A} = (Ax\,(x, y, z, t),\ 0,\ 0)$. Using the apparatus shown in [4–9], Eq. (1), we obtain an effective equation for the vector potential:

$$\frac{\partial^2 A_z}{\partial x^2} + \frac{\partial^2 A_z}{\partial y^2} + \frac{\partial^2 A_z}{\partial z^2} - \frac{1}{c}\frac{\partial^2 A_z}{\partial t^2} + \frac{q}{\pi\hbar}\sum_m c_m \sin\left(\frac{maq}{c}A_z(t)\right) = 0 \qquad (3)$$

where F0 is the equilibrium Fermi distribution function, q is the elementary electron charge, q0 is the pulse on the boundary of the Brillouin zone $q_0 = \dfrac{2\pi\hbar}{3b}$. The coefficients $a_{ms}$ are determined by Fourier series expansion of velocity component $v_x\,(s, p)$ obtained from (1) as $v_x = \partial E(p)/\partial p_x$.

These simplifications have been made: (a) interband transitions are not counted. This assumption imposes a restriction on the maximum frequency of laser pulses, which for the CNT is in the near infrared [11]. (b) the solution for the field component $A_x(t)$ is in the class of rapidly decreasing functions. (c) the relaxation time $\tau$ is large enough for the common durations of ultrashort laser pulses.

Equation (3) was solved numerically using the direct difference scheme "cross" [12]. Steps to time and coordinate were determined from the standard conditions of stability and decreased successively in 2 times as long as the solution is not changed in the eighth digit after dot. These calculations for the one-dimensional case are presented in [4–9]. The evolution of the initial conditions considered in time (the variable $t$). The initial pulse profile has a Gaussian form:

$$E(x,y,z,t) = E_{max} \exp\left(-\frac{(x-x_0)^2}{\mu_x}\right)\exp\left(-\frac{(y-y_0)^2}{\mu_y}\right)\exp\left(-\frac{(z-z_0)^2}{\mu_z}\right)$$

where $E_{max}$ is the maximum field amplitude, $x_0$ is the coordinate of the maximum intensity the axis of $x$, $y_0$ is the coordinate of the maximum intensity the axis of $y$, $z_0$ is the coordinate of the maximum intensity the axis of $z$, $\mu_x$, $\mu_y$ and $\mu_z$ are the parameters defining pulse width to the coordinates $x$, $y$, and $z$, respectively.

## 4.3  RESULTS AND DISCUSSION

This section focuses on the results of the impulse, which localized in three spatial dimensions. In the evolution of ultrashort pulse periodically split into two pulses and then combined into one, while the amplitude of the pulses is changed. Figures 2 and 3 shows the results of numerical simulation of the pulse evolution in the three-dimensional CNTs array such as (8.0) and (20.0), respectively.

Figures 2 and 3 shows the distribution of intensities $I_x(x, y, z, t) = E_x^2(x, y, z, t)$ at different times. Analysis of the data graphs show that the stable propagation of pulses in the array of CNTs is possible, which are localized in three dimensions. These pulses are often called in literature as "light bullets" [15, 16]. The interesting effect of a periodic partition of pulse maximum into two maxima, and their subsequent merger into a single maximum was diskovered. The pulse shape varies significantly during the evolution of the pulse. We consider that this effect is akin to the dynamics of internal modes [13,14] and consists in the excitation of internal vibration modes of the "light bullet" when passing a pulse through the array of CNTs, which leads to periodic separation of the pulse maximum and their interference with the original "bullet." This fact suggests that there may be bumeron analogs in the other highly nonlinear three-dimensional systems.

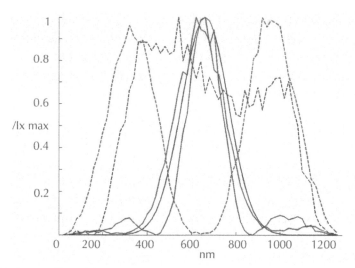

**FIGURE 2**   Profiles of the intensities $I_x/I_{xmax}$ bullets in the cross section parallel to the plane *ZOY* and passing through a maximum pulse propagating in the array of CNTs such as (8.0). The ordinate shows the ratio $I_x/I_{xmax}$. Type of pulse: 1 – initial pulse, 2 – pulse after $t = 0.8 \cdot 10^{-12} s$, 3 – pulse after $t = 1.6 \cdot 10^{-12} s$, 4 – pulse after $t = 2.4 \cdot 10^{-12} s$, 5 – pulse after $t = 2.9 \cdot 10^{-12} s$.

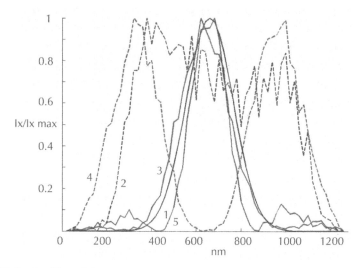

**FIGURE 3**   Profiles of the intensities $I_x/I_{xmax}$ bullets in the cross section parallel to the plane ZOY and passing through a maximum pulse propagating in the array CNT type (20.0). The ordinate shows the ratio $I_x/I_{xmax}$. Type of pulse: 1 – initial pulse, 2 – pulse after $t = 0.8 \cdot 10^{-12} s$, 3 – pulse after $t = 1.6 \cdot 10^{-12} s$, 4 – pulse after $t = 2.4 \cdot 10^{-12} s$, 5 – pulse after $t = 2.9 \cdot 10^{-12} s$.

## 4.4 CONCLUSION

We formulate the main results in conclusion.

1. The model and the effective equation describing the dynamics of an ultrashort pulse laser beams in the CNT are obtained. The approximations, which used in this model, were formulated.
2. Numerical calculations showed that in three-dimensional case are possible stable nonlinear waves that are localized in three directions of light pulses, which are analogs of "light bullets."
3. The periodic separation of peak pulse into two pulses and subsequent integration occur in the case of scattering when light pulse passing through the array of CNTs. The pulse shape at the same time varies considerably.

## KEYWORDS

- **Brillouin zone**
- **light bullets**
- **Ultrashort optical pulse**
- **zigzag carbon nanotubes**

## REFERENCES

1. Zheltikov, A. M., UFN. **2007,** *177(7),* 737.
2. Maksimenko, S. A.; Slepyan, G.Ya. Handbook of nanotechnology. Nanometer structure: theory, modeling, and simulation. SPIE Press. Bellingham, **2004,** 145.
3. Harris, P. Carbon nanotubes and related structures. *New materials of the XXI century.* M.: Technosphere, **2003,** 336.
4. Belonenko, M. B.; Demushkina, E. V.; Lebedev, N. G., *J. Russian Laser Res.* **2006,** *27,* 457.
5. Belonenko, M. B.; Demushkina, E. V.; Lebedev, N. G., Math. Academy of Sciences. *Ser. Phys.* **2008,** *72.* 28.
6. Belonenko, M. B.; Demushkina, E. V.; Lebedev, N. G., *Solid State.* **2008,** *50,* 367.
7. Belonenko, M. B.; Demushkina, E. V.; Lebedev, N. G., Math. Academy of Sciences. *Ser. Phys.* **2008,** *72.* 711.

8. Belonenko, M. B.; Demushkina, E. V.; Lebedev, N. G., JTF. **2008,** *78,* C. 1.

9. Belonenko, M. B.; Demushkina, E. V.; Lebedev, N. G., Chem. Physics. **2008,** *27*, 97.

10. Ilyinsky Y.; Keldysh L. The interaction of electromagnetic radiation with matter. Moscow: Moscow University Press, **1989,** 304.

11. Epstein, E. M., Solid State. **1976,** *19,* 3456.

12. Bakhvalov, N. S. Numerical methods (analysis, algebra, ordinary differential equations). Moscow: Nauka, **1975,** 630.

13. Serf, P.; Popov, V. O.; Rozanov, N. N., Opt. **2000,** *89*, 964.

14. Wittenberg A.; Castle; Popov O.; Rozanov, N. N., Opt. **2002,** *92*, 603.

# SECTION III: A NOTE ON GRADIENTLY ANISOTROPIC CONDUCTING AND MAGNETIC POLYMER COMPOSITES

J. ANELI, L. NADAREISHVILI, G. MAMNIASHVILI, A. AKHALKATSI, and G. ZAIKOV

## 4.1 INTRODUCTION

It is well known that there are several methods for obtaining of materials with anisotropic properties by chemical methods (copolymerization, polymer-analogous transformation, radiation-chemical modifications, etc.) [1,2]. At present for obtaining of such structures one of the best methods is the orientation of polymer films in the definite direction and environment conditions. It is known also that at stretching of thermoplastic polymers above glass temperature the material in orientation state is formed. Such polymers are characterized with mono-axis crystal symmetry. In this state the principal direction of macromolecules coincides with the direction of stretching. If the polymer filled with different dispersive fillers, particularly with electric conductive and magnetic materials (metal powders, graphite, carbon black), the particles distribution of lasts interacting with macromolecules transform from chaotic state to orientation one. The change of polymer microstructure in this case entirely defines the material properties [3,4].

In the presented work, the character of change of electric conductivity and magnetic properties of composites films based on polyvinyl alcohol, graphite and nickel at their mechanical stretching has been investigated.

## 4.2   EXPERIMENTAL METHODS

The films were prepared with using of following technology:

To the fine dispersive graphite or nickel suspensions in polyvinyl-alcohol water solution were prepared. The mixture was filtrated and the film was formed on the dryer table.

The specific volumetric electric resistance of the polymer films obtained with using of described above method was changed in the interval 10–50 kOhm.sm. The selection of such interval of the composite resistance was dictated by preliminary selection of conducting composites effectively reacted on the mechanical deformations. The composites contained the magnetic filler are characterized simultaneously both electrical and magnetic properties.

The experiments were carried out on the basis of polymer composite films with rectangle and trapezoidal shape. The width of films was no more than 0.2 mm. The deviation of the resistance of any local region of the film was no more than 10%. These films were fixed in special clamps, placed to the heater and were stretched on 50–150% at rate 50 cm/min and temperatures 100–120°C. Stretching was conducted for rectangle form sample along big side and for trapezoidal sample in parallel to bases direction (Figs. 1 and 2).

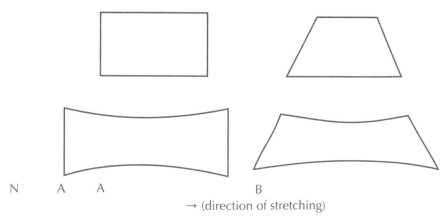

N       A      A                              B

→ (direction of stretching)

**FIGURE 1**   Rectangle (A) and trapezoidal (B) shape films before (top) and after (bottom) stretching along big sides.

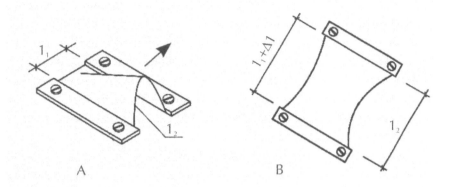

**FIGURE 2**   The scheme of orientation of trapezoidal shape film. (A) before stretching, (B) after stretching.

After stretching of the deformed films local ohmic resistances were measured. First of all, it was necessary to mark the film with square grid. In our case the length of square side was equal to 5 mm. The local resistances were measured by using of twin needles after touching them to the film. The measuring of resistance of elementary cages were performed several times and than the average significances of resistances were calculated.

## 4.3   RESULTS AND DISCUSSION

As a result of measuring of local resistances of oriented rectangle shape samples it was established that maximum change of this parameter was noticed along symmetry axis taken parallel to longest side of the rectangle. This change has an extreme character (the maximum is at the central part of the film) and its full shape has Gaussian form. Fig. 3 shows that the amount of maximum of the local resistances depends on the value of stretching. This result is in good agreement with the known conception on the mechanism of conductivity of conductive polymer composites [3]. The analogical character has the same dependence for films in rectangle to stretching direction. It has an analogical character of the same dependence, although this dependence is some-what weaker.

It was interesting to establish the character of considered    a b o v e functional dependences of the local conductance on the concentration of powder electric conducted particles. Fig. 4 shows that the increase of filler concentration leads to decreasing of resistance change at stretching. This phenomenon may be described by increasing of degree of reservation of electric conductive channels [4].

The geometry of distribution of local resistances in case of trapezoidal shape films is more complex, although in this case it is possible to describe definite lows of this problem.

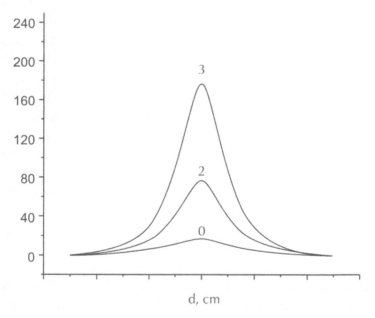

**FIGURE 3** Dependence of local resistances of polymer film on the value of stretching parallel to long side of the rectangle: 1–50%, 2–100%, 3–150%. Central ordinate passes through the coordinate "0.".

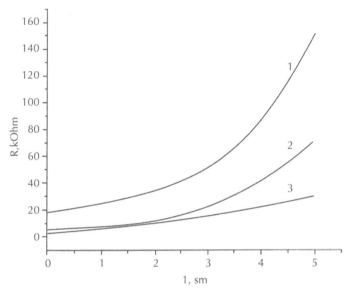

**FIGURE 4** Dependence of local resistances of films on the stretching degree for composites based on PVA containing 30 (1), 40 (2) and 50 (3) mas.% of graphite powder (curves correspond to left half side of central strip of the film (see Fig. 3, b)

a)

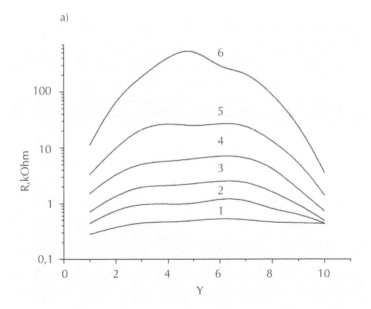

**FIGURE 5** *(Continued)*

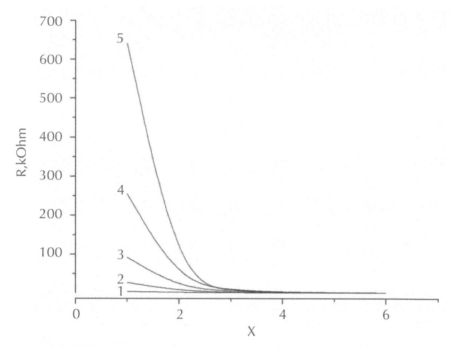

**FIGURE 5** Dependences of local resistances on the film coordinates in the stretched (150%) trapezoidal shape films on the basis of PVA with graphite perpendicular to stretching direction from big base to small one (a) and parallel to stretching direction from small base to big one (b). The numbers on the curves indicate the numbers of stripes in perpendicular (a) and parallel (b) to stretching direction. The curve number 5 of Fig. 5b corresponds to central strip of trapezoidal film and others – to side from it (4-3-2-1). The asymmetry of the curves on the Fig. 5a is due to inhomogeneous distribution of the filler particles in polymer matrix to some extent.

The following series was fulfilled on the trapezium of the same shape, but with one difference only small base and nearest to it region was stretched. Here was created the mechanical stretch gradient in perpendicular to base direction (Fig. 5). This gradient was increased from zero at big base and was ended at more stressed small base with maximum.

Similar dependencies are observed also for local magnetic characteristics of gradient anisotropic magnetic polymeric composites with magnetic powder fillers like nickel powder.

The films from magnetic polymer composites, fabricated by the above described method, were subjected to the similar stretching procedure as it was previously described.

The resulting distribution of magnetic particles density in polymer composites was recorded by the method of LC-generator similar to the one used by us in work [5] for NMR detection in europium garnet at low temperatures.

It is also interesting to note that a similar method was used for the first precision determination of the magnetic field penetration length in super-conductors [6].

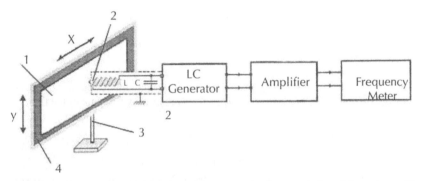

**FIGURE 6**   Scheme of the measuring of the magnetic characteristics of the polymer films. 1 – Sample; 2 – ferrite tip, 3 – support, 4 – frame.

The experimental set-up is presented in Fig. 6. In the inductive coil of the resonance contour of LC-generator it is placed cylindrical tipped ferrite rod used as probe. The investigated rectangular shape magnetic polymer composite film is displaced relatively the immovable ferrite tips. The scanning of the film surface is realized along the previously marked net contour (Fig. 6).

The change of magnetic particle concentration causes the change of inductance $\delta L$ of the resonance contour of LC-generator resulting in the frequency displacement of LC-generator $\delta f$ related with $\delta L$ by relation $\delta f/f = 1/2\delta L/L$. This frequency displacement could be precision measured what stipulates the high sensitivity of the method.

At the natural frequency of used LC-generator near ~2 MHz the observed range of the frequency change δf was about был ~1000 Hz at the precision of the frequency measurement ~1 Hz.

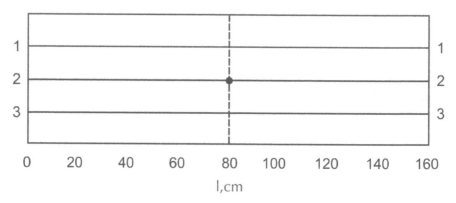

**FIGURE 7**  Scheme of ferromagnetic film for investigation of the local magnetic susceptibility along directions 1–1, 2–2 and 3–3; dotted line – the middle of the film.

Measurements of the dependences F – l were provided with using of the ferromagnetic film, the scheme of which is presented in the Fig. 7.

In Fig. 8 it is presented the results of measurements of generator frequency change along contour lines 1–1, 2–2 and 3–3 of network marked on film surface. Analogical results are obtained at measuring of magnetic susceptibility of film along rectangular to 1–1, 2–2 and 3–3 directions.

The presented results show that for magnetic polymer composite films it is observed also dependences similar to those ones for conducting polymer composites, because frequency is reduced when magnetic particle concentration and, consequently, local susceptibility and inductance increases.

Results of detailed measurements will be published elsewhere.

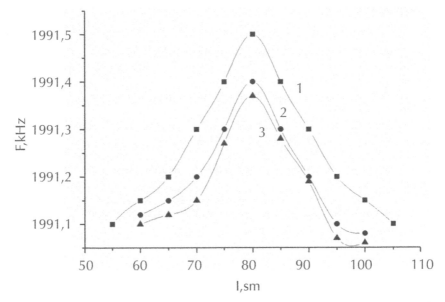

**FIGURE 0**  Dependences of the oscillation frequency of the ferrite sensor on the coordinates of film made from polymer composite along directions 2–2 (1), 1–1 (2), 3–3 (3).

## 4.4  CONCLUSION

The experiments described above open the perspectives in the field of creation of the films with desirable anisotropy of electric conductance and magnetic properties. The perspective is in application of materials described above for creation of the so-called printed schemes. In electronics these films will be useful for preparing of some functional microschemes.

## KEYWORDS

- **ferromagnetic film**
- **mono-axis crystal**
- **polyvinyl alcohol**
- **trapezoidal shape film**

## REFERENCES

1. Polymers and polymeric materials for the fiber and gradient optics / VSP (Utrecht-Boston-Köln-Tokyo), **2002,** (With Lekishvili, N.; Nadareishvili, L., et al.), 230 (Monograph).
2. Nadareishvili, L. et al. GB-optics: A new direction of gradient optics, *J. Appl. Pol. Sci.* **2004,** *91,* 489–493.
3. Aneli, J. N.; Khananashvil, L. M.; Zaikov, G. E. Structuring and conductivity of polymer composites. *Nova Sci. Publ. N. Y.* **1998,** 326.
4. Aneli, J. N.; Khananashvil, L. M.; Zaikov, G. E. Effects of mechanical deformations on structurization and electric conductivity of polymer composites. *J. Appl. Pol. Sci.* **1999,** *74,* 601–621.
5. Pavlov, G. D.; Chekmarev, V. P.; Mamniashvili, G. I.; Gavrilko, S. I. "Method of NMR recording in magnetoordering materials." USSR Patent (27989) **1988**.
6. Schawlow, A. L.; Devlin, G. E. *Phys. Rev.* **1959,** *113(1),* 120–126.

# CHAPTER 5

# STABILIZATION PROCESS OF PAN NANOFIBERS

S. RAFIEI, B. NOROOZI, SH. ARBAB, and A. K. HAGHI

## CONTENTS

## 5.1   INTRODUCTION

Carbon fibers are one of the most famous kinds of carbonaceous materials due to their special characteristics such as high conductivity, strength and modulus [1, 2]. They are typically produced either by pyrolyzing fibers spun from an organic precursor (e.g., Polyacrylonitrile (PAN), polyimide, mesophase pitch,…) or by chemical vapor decomposition (CVD) [3 4]. Since the performance of carbon fiber composites highly depends on its precursor and the desired precursor should have high carbon content, high molecular weight and also high degree of molecular orientations, PAN based carbon fiber seems to satisfy all of these expectations [2, 5–7]. PAN based carbon nanofibers are more common because they are low-cost, continuous, and easy to be aligned and assembled. They also possess high degrees of macromolecular orientations induced by the high stretching ratios (up to 105) under the electrical force [8]. Due to their high mechanical, thermal and electrical performances PAN based carbon nanofibers can be applied promisingly in making super-capacitors, composites, etc. [9].

Carbon nanofibers derived from electrospun PAN are long (centimeter long nanofibers are fabricated routinely), continuous and relatively well aligned, hence they are ideal for strengthening, stiffening and toughening of polymers [10]. PAN homopolymer has been hardly used because of its poor properties and processability, so the monomers such as acrylic acid (AA), methacrylic acid (MAA) and methylmethacrylate (MMA) have been widely added to PAN monomer to form high performance carbon fibers [11].

Electrospinning method uses electrostatic forces to spin ultrafine fibers from precursor solution. The common spinning techniques can produce micro scale carbon fibers (diameter> 5 μm), while by employing electrospinning of precursor solution, especially PAN, it is possible to get carbon nanofibers in nanoscale [7, 9, 12–14]. PAN precursor fibers and nanofibers need to be thermally stabilized prior to pyrolysis process to obtain high quality carbon fibers. PAN nanofibers are converted into carbon nanofibers by the sequenced processes of stabilization, carbonization and graphitization [10]. The stabilization is intended to prevent melting or fusion of the fiber, to avoid excessive volatization of elemental carbon in the subsequent carbonization step and thereby to maximize the ultimate carbon yield from the fiber precursor. The chemistry of the stabilization process is complex, but generally consists of cyclization of the nitrile groups ($-C\equiv N$) and cross-linking of the chain molecules in the form of $-C=N-C=N-$. In this process a lower temperature treatment (180–300°C) in air (usually oxygen containing atmosphere) leads to the formation of an aromatic structure. Dehydrogenation, aromatization, oxidation and cross-linking occurs as a result of the conversion of $C\equiv N$ bonds to $C=N$ bonds, then fully aromatic cyclized ladder type structure forms. This cyclized structure is stable toward heat and is converted to turbo-stratic carbon [2, 8, 14–17]. The physical changes include color, crystallinity, entropy, density, and tensile strength; while chemical changes include nitrile polymerization, chain scission, evolution of volatile gasses ($NH_3$, $CH_4$, $CO$, $CO_2$, $H_2O$, and so on), and formation of carbonyl, carboxyl, and peroxide groups [18]. Also, it has been reported that during stabilization, $-CH_2$ and $-C\equiv N$ groups disappear while $C=C$, $C=N$ and $=C-H$ groups form. At the same time the color of precursor fiber changes gradually and finally turns black when carbonized [2, 6, 19].

This step is the most important part of carbon fibers and nanofibers formation and has been attracted a lot of attentions [6, 14, 20,]. It has been shown that thermal stabilization conditions have considerable effect on the properties of the resulting carbon fiber [14]. It is necessary to choose the optimum conditions of temperature, heating rate and oxidation duration. Studies showed that optimal stabilization conditions lead to high modulus carbon fibers. Too low temperatures lead to slow reactions and incomplete stabilization, whereas too high temperatures can fuse or even burn the fibers [6, 9].

The physical, chemical and structural changes created in stabilization can be detected by various analyzing techniques include X-ray diffraction (XRD), Differential scanning calorimetery (DSC) and Fourier transform infrared (FTIR) spectroscopy, which can be considered as standard evaluation techniques for the extent of stabilization [14, 21, 22]. In this study, different methods are employed to measure stabilization index of electrospun PAN nanofibers after heat treating to obtain the desired conditions for stabilizing temperature, dwell time and heating rate.

## 5.2 EXPERIMENTAL METHODS

### 5.2.1 MATERIALS

Poly(acrylonitrile-rar-methylacrylate) (94.6 wt%) with molecular weight of $1,00,000–1\times10^5$ g/mol received from Polyacryl Co. (Isfahan, Iran). Dimethylformamide (DMF) was purchased from Merck Company as appropriate solvent due to its solubility parameters [23] of PAN powder dissolving.

### 5.2.2 PAN NANOFIBERS PREPARATION

Accurate scaled mass of PAN powder dried and stored in dessicator for 24 hours was dissolved in DMF. The solutions with concentration ranges of 8–13 wt% were prepared by using magnetic stirrer for 1 day. Then each solution was fed into the electrospinning set up by a 2 ml syringe pump with standard pumping rate. The Gamma high voltage was used to supply applicable voltage of 7–10 KV in the spinning distances of 15 cm for all samples. Scanning electron microscopy (SEM) was used to select the best electrospinning conditions according to the appearances, uniformity and diameters of electrospun nanofibers. The electrospinning set up used in this work can be found elsewhere [24].

### 5.2.3 THERMAL STABILIZATION OF PAN NANOFIBERS

Stabilization of the samples was carried out under air atmosphere in a chamber furnace (Nabertherm Controller Co.) using different temperatures and dwell times ranging from 180 to 270°C and 15–120 min, respectively. All runs were in discontinuous mode. Heating rate also varied from 1 to 4° C to obtain the best results. The range of temperatures was chosen according to the thermal behavior of PAN nanofibers got from DSC analysis.

## 5.2.4   CHARACTERIZATION OF NANOFIBERS

Philips Scanning electron microscope (XL-30, Poland) was employed to get micrograph from stabilized and untreated samples. FTIR (Nexus 670) from Nicolet Instrument Corporation was used to detect the chemical transformation as well as measuring the extent of reaction (EOR) after heat stabilization. DSC measurements were carried out using DSC 302 calorimeter (BAHR Thermo Analyzes Co.). All samples were weighed in 7 mg in aluminum pans using alumina reference. X ray diffractometer (Philips 1140/90, Poland) with Ni filtered Co Kα radiation ($\lambda$=1.785 Å) was used to obtain XRD patterns of PAN fiber and the fibers stabilized in different conditions. Each sample had a weight around 4 mg. The fibers were fixed straightly on the supporting glass. The scanning rate is 4°/min with a scanning step of 0.02.

## 5.3   RESULT AND DISCUSSION

The PAN nanofiber and their stabilized form were evaluated using SEM, FTIR, DSC and X-ray analysis. SEM photographs showed that the best applied voltage to obtain the uniform electrospun PAN nanofiber webs having no beads in their structure using mentioned concentrations range was 8 KV. This is an important key factor in the subsequent heat treatment of the web for carbon and activated carbon nanofiber production. Table 1 shows the average diameters of produced PAN nanofibers in concentration range of 8–13% under constant voltage, and spinning distance of 8KV, and 15 cm, respectively.

**TABLE 1**   Average diameter of prepared PAN nanofibers, (Voltage: 8KV, pumping rate: 2 μl/min and spinning distance: 15 cm).

| Concentration (wt%) | 8 | 10 | 11 | 13 |
|---|---|---|---|---|
| Average diameter (nm) | 78.2 | 121.4 | 367.5 | 1100.0 |

Studies showed that PAN nanofibers lose about 50% of their diameters during stabilization and carbonization processes [20, 25], so concentrations of 8 and 10% seem to be inappropriate for this study because of their instability during the heating process. This was revealed in selected experiments. As it is seen in Table 1 the concentration of 13% will not provide desirable diameter due to its micro scale. Figure 1 shows SEM pictures of PAN nanofibers spun using 11% solution.

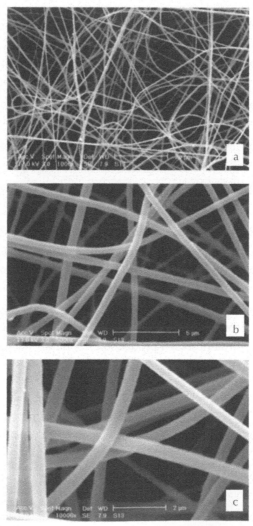

**FIGURE 1**    SEM micrograph of PAN nanofiber in optimum conditions. (a: 1,000′, b: 5,000′, c: 10,000′) (Voltage: 8 KV, pumping rate: 2 μl/min, distance: 15 cm).

Comparing the FTIR spectra of untreated and stabilized PAN nanofibers in Fig. 2, the effects of temperature and duration time on chemical bond changes can be concluded. During stabilization, the peak at 2,240 cm$^{-1}$ related to C≡N bond decreases intensely, indicating nitrile groups disappearing. On the other hands, new peaks around 1,600 cm$^{-1}$ and 1,400 cm$^{-1}$ engender that are related to the formation of C=N and C=C bonds, reveals improving the creation of aromatic and ladder like structures [6, 25]. A common way to assess the extent of stabilization reaction is using Eq. (1):

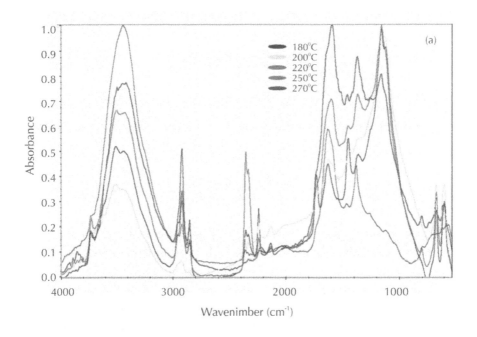

FIGURE 2   (Continued)

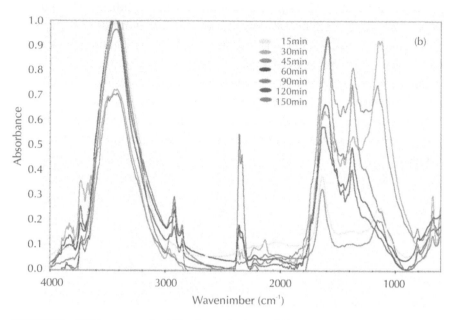

**FIGURE 2**   FTIR spectra of stabilized samples in (a) different temperatures and (b) dwell times.

$$EOR\% = \frac{I_{1600}}{I_{2240} + I_{1600}} \tag{1}$$

Where $I_{2240}$ and $I_{1600}$ show the intensity of peaks related to C≡N and C=N bonds, respectively. Table 2 shows the effect of temperature on the EOR. Data indicate that increasing temperature has an extraordinary effect on the extent of reaction and weight loss percentage (WL%). Temperatures in the range of 250–280°C seem to afford good results but due to their weight loss, 280°C is not recommended for stabilization process. Dwell time as presented in Table 3, has the same effect, but with less intensity, on EOR and WL%. The dwell time between 60–120 min is optimum due to calculated EOR%. The influence of heating rate is negligible. Making a brittle and cracky appearance, heating rates more than 2°C/min do not provide proper conditions for stabilization process.

**TABLE 2**   Effects of temperature on EOR and WL% of stabilized nanofibers (constant time and heating rate of 1h, 1°C/min).

| Stabilization temperature (°C) | 180 | 200 | 220 | 250 | 270 | 280 |
|---|---|---|---|---|---|---|
| EOR% | 62.14 | 67.48 | 76.08 | 89.52 | 91.28 | 92.21 |
| Weight loss% | 18.0 | 22.3 | 25.8 | 30.0 | 32.1 | 38.4 |

**TABLE 3**   Effects of time on EOR and WL% of stabilized nanofibers (constant temperature and heating rate of 250°C, 1°C/min).

| Stabilization time (min) | 15 | 30 | 45 | 60 | 90 | 120 | 150 |
|---|---|---|---|---|---|---|---|
| EOR% | 83.3 | 84.4 | 84.8 | 89.52 | 91.8 | 92.03 | 92.08 |
| Weight loss% | 19.2 | 25.6 | 26.8 | 28.1 | 30.5 | 32.4 | 37.9 |

A different method for evaluating the stabilization extent is to study thermal behavior of untreated and stabilized PAN nanofibers using DSC analysis [19] in certain heating rate (in this study 10°C/min). As mentioned, the DSC curve of PAN nanofibers has two exothermic peaks, the first one is related to the cyclization reactions and the latter confirms oxidation processes. Thermally stabilized nanofiber shows wider exothermic peaks with lower heights. Figure 3 shows DSC curves of nanofibers stabilized in different conditions. The conditions (times and temperatures) are selected according to higher calculated EOR.

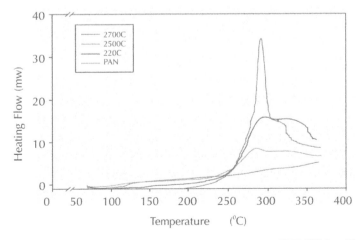

**FIGURE 3**   *(Continued)*

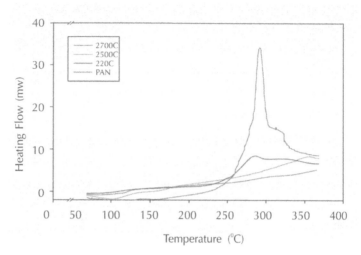

**FIGURE 3**   DSC curves of stabilized samples in different temperatures (up) and dwell times (down).

The curves show that during the stabilization, the intensity of exothermic peaks are reduced and peaks altered to the lower temperature, indicating that cyclization of the structure occurred more rapidly by increasing stabilization time and temperature. The plot area of the curves, representing the enthalpy of reaction, was decreased by completing the process. The reduction of enthalpies can be used to measure stabilization yield, by Eq. (2) [26].

$$SI\% = \frac{\Delta H_{ox}}{(1-\Delta H_0)} \times 100 \tag{2}$$

Where SI% is stabilization index percentage, $\Delta H_{ox}$ is the enthalpy of oxidized PAN nanofibers and $\Delta H_0$ is the enthalpy of raw PAN nanofibers.

Calculated SI presenting in Tables 4 and 5, shows consistency to the related FTIRs EOR%. DSC analysis confirms that the most important effective parameter in stabilization extent is temperature.

**TABLE 4** Enthalpies and stabilization indexes of stabilized samples in different temperatures (constant time of 1h, and heating rate of 1°C/min).

| Samples | PAN | Stabilized PAN (Temperature (°C)) | | |
|---------|-----|-----|-----|-----|
| | | 200 | 250 | 270 |
| ΔH (J) | 180.8 | 63.2 | 17.4 | 11.1 |
| SI% | – | 65.0 | 90.4 | 93.8 |

**TABLE 5** Enthalpies and stabilization indexes of stabilized samples in different dwell times (constant temperature of 270°C, and heating rate of 1°C/min).

| Samples | PAN | Stabilized PAN (Duration time (h)) | | |
|---------|-----|-----|-----|-----|
| | | 1 | 2 | 3 |
| DH (J) | 180.8 | 11.1 | 10.8 | 9.1 |
| SI% | NA | 93.8 | 94.0 | 95.1 |

The last method has been used to measure the stabilization index was utilization of XRD patterns of untreated and stabilized nanofibers. XRD has usually been used to study the crystalline structures changes during thermal stabilization by many researchers [11, 27]. Figure 4 shows the XRD patterns of PAN precursor fibers and the fibers stabilized at various temperatures. It's obvious that the main diffraction peak for PAN fibers occurs at around $2\theta=17°$, which is corresponding to plane (100) of a hexagonal structure. Another peak is around $2\theta=29°$ that is related to plane (110) [28]. As mentioned before, when the PAN nanofiber is heated above 180°C in the presence of oxygen, $C\equiv N$ bonds are converted into $C=N$ bonds as the ladder polymer is formed. In this time, a new diffraction peak in $2\theta=25°$ comes to exist; corresponding to a sheet-like structure ladder polymer. The percentage conversion of $C\equiv N$ to $C=N$ groups depends on the extent of stabilization and can be attributed to decreasing the intensity of the peak in $2\theta=17°$ and forming a broad peak in $2\theta=25°$ [22]. So the stabilization index can be defined as Eq. (3).

$$SI\% = \frac{I_0 - I_i}{I_0} \times 100 \tag{3}$$

Where $I_0$ is the intensity of the diffraction peak of the original PAN nano-fibers in $2\theta = 17°$ and $I_i$ is the intensity of the same peak in the stabilized nanofibers. The SI value increases with temperature or heat-treatment time during stabilization. Figure 4 shows XRD patterns of samples in different stabilization temperatures. As Table 6 shows, after stabilization in $270°$, the yield of the reaction gets to 90% and the process is nearly accomplished. Evaluation methods of stabilization index that is discussed, is one of the most accurate technique that can be employed to optimize pre-oxidation parameters. The results obtained are similar to previous methods datas.

**TABLE 6**   AI and SI values of the stabilized nanofibers in different Temperature.

| Stabilized PAN (Temperature (°C)) | | | PAN | Samples |
|---|---|---|---|---|
| 270 | 250 | 200 | | |
| 90.0 | 83.4 | 56.1 | NA | SI% |
| 62.3 | 50.0 | 22.0 | NA | AI% |

Aromatization Index (AI) is a useful parameter indicating the content of aromatic structures in stabilized polymer can be obtained from XRD patterns. AI is calculated by Eq. (4) [29]:

$$SI\% = \frac{I_a}{I_a - I_p} \times 100 \tag{4}$$

Where, $I_a$ and $I_p$ are the intensity of a new peak ($2\theta = 25°$) and the main peak at $2\theta = 17°$.

Ko et al. [29] modified this testing method and recommended that the AI value can be used to check the stabilization process and estimate the ladder structure of the polymer. A higher percentage conversion (AI value) means that the stabilized fibers have more ladder characteristics. Table 6 shows AI and SI values of the stabilized nanofibers. As a result the higher stabilization temperature yields a stabilized polymer with greater ladder content. A suitable AI value of stabilized fibers for activating is calculated about 50–60%.

## 5.4  CONCLUSION

The quality of the resulting carbon fibers depends strongly upon the degree of stabilization process during carbon nanofiber production. Hence it is important to evaluate the stabilization accurately and rapidly so as to provide information for optimizing technology parameters such as dwell time, temperature and heating rate. In this study untreated and stabilized PAN samples were evaluated using FTIR, DSC and XRD analysis. All results indicated that both dwell time and temperature of stabilization process increase the extent of reaction, but the influence of the temperature is more obvious. FTIR spectra of stabilized samples showed that the extent of the reaction is positively affected by increasing time and temperatures. In all experiments heating rate influence was negligible. In DSC curves of the samples, total enthalpy decreased by increasing the mentioned parameters, revealed improving the stabilization index. XRD patterns also used to evaluate stabilization yield and AI value of stabilized fibers for activation was calculated about 50–60%. Finally, according to the assessment of all applied methods, consistency was observed among the obtained results to propose the best process conditions.

## KEYWORDS

- **aromatization index**
- **differential scanning calorimetery**
- **dimethylformamide**
- **Fourier transform infrared spectroscopy**
- **scanning electron microscopy**

## REFERENCES

1. Fitzer, E. Carbon Fibers and Their Composites. Springer-Varlang, Berlin, **1986**.
2. Yusof, N.l Ismail, A. F. Post spinning and pyrolysis processes of polyacrylonitrile (PAN)-based carbon fiber and activated carbon fiber: a review. *J. Analytical and Applied Pyrolysis* **2012**, *93*, 1–3.
3. Morgan, P.; Raton, B. In *Carbon Fiber and Their Composites*. Taylor and Francis: New York. **2005**,
4. Endo, M. Grow carbon fibers in the vapor phase. *Chemtech* **1998**, *18*, 568–576.
5. Kima, J.; Ganapathya, H.; Hongb, S.; Lim, Y. Preparation of polyacrylonitrile nanofibers as a precursor of carbon nanofibers by supercritical fluid process. *J. Supercritical Fluids* **2008**, 47: 103–107.
6. Esrafilzadeh, D.; Morshed, M.; Tavanai, H. An investigation on the stabilization of special polyacrylonitrile nanofibers as carbon or activated carbon nanofiber precursor. *Synthetic Metals* **2009**, *159*, 267–272.
7. Lai, C.; Zhong, G.; Yue, Z.; Chen, G.; Zhang, L.; Vakili, A.; Wang, Y.; Zhu, L, Liu, J.; Fong, H. Investigation of postspinning stretching process on morphological, structural, and mechanical properties of electrospun polyacrylonitrile copolymer nanofibers. *Polymer* **2011**, *52*, 519–528.
8. Wu, M.; Wang, Q.; Li, K.; Wu, Y.; Liu, H. Optimization of stabilization conditions for electrospun polyacrylonitrile nanofibers. P*olymer Degradation and Stability* **2012**, *97*, 1511–1519.
9. Nataraj, S. K.; Yang, K. S. Aminabhavi ™ Polyacrylonitrile-based nanofibers: a state-of-the-art review. *Progress in Polymer Science* **2011**, *37*, 487–513.
10. Arshad, S.; Naraghi, M.; Chasiotis, I. Strong carbon nanofibers from electrospun polyacrylonitrile. Carbon **2011**, *49*, 1710–1719.
11. Mukesh, K. J.; Balasubramanian M Conversion of Acrylonitrire Based Percursor to Carbon Fibers. J. Material Science **1987**, *22*, 301–310.
12. Bhardwaj, N.; Kundu, S. C. Electrospinning: a fascinating fiber fabrication technique. Biotechnology Advances **2010**, *28,* 325–347.
13. Reneker, D.; Chun, I. Nanometer diameter fibers of polymer, produced by electrospinning. Nanotechnology **1996**, *7*, 216.
14. Fazlitdinova, G.; Tyumentsev, A.; Podkopayev, A.; Shveikin, P. Changes of polyacrylonitrile fiber fine structure during thermal stabilization. J. Material Science **2010**, *45*, 3998–4005.
15. Panapoy, M.; Dankeaw, A.; Ksapabutr, B. Electrical Conductivity of PAN-based Carbon Nanofibers Prepared by Electrospinning Method. Thammasat International J. Science and Technology **2008**, *13*, 88–93.
16. Kim, C.; Park, S. H.; Cho, J. I.; Lee DY, Park, T. J.; Lee, W. J.; Yang, K. S. Raman spectroscopic evaluation of polyacrylonitrile-based carbon nanofibers prepared by electrospinning. *J. Raman Spectroscopy* **2004**, *35*, 928–933.
17. Yusof, N.; Ismail, A. F. Post spinning and pyrolysis processes of polyacrylonitrile (PAN)-based carbon fiber and activated carbon fiber: a review. *J. Analytical and Applied Pyrolysis* **2012**, *93,* 1–13.

18. Wu, G.; Lu, C.; Ling, L.; Lu, Y. Comparative investigation on the thermal degradation and stabilization of carbon fiber precursors. *Polymer Bulletin* **2009,** *62,* 667–678.
19. Belyaeva, S. S.; Arkhangelsky, I.; Makarenkob, V. Non-isothermal kinetic analysis of oxidative stabilization processes in PAN fibers. *Thermochimica Acta* **2010,** *507–508,* 9–14.
20. Wangxi, Z.; Jie, L.; Gang, W. Evolution of structure and properties of PAN precursors during their conversion to carbon fibers. *Carbon* **2003,** *41,* 2805–2812.
21. Jie, L.; Wangxi, Z. Structural Changes during the Thermal Stabilization ofModified and Original Polyacrylonitrile Precursors. *J Appl Polym Sci* **2005,** *97,* 2047–2053.
22. Yu, M.; Bai, Y.; Wang, C.; Xu, Y. Guo A new method for the evaluation of stabilization index of polyacrylonitrile fibers. *Materials Letters* **2007,** *61,* 2292–2294.
23. Heikkilä, P.; Harlin A Electrospinning of polyacrylonitrile (PAN) solution: effect of conductive additive and filler on the process. *Express Polymer Letters* **2009,** *3,* 437–445.
24. Ziabari, M. Mottaghitalab, V.; Haghi, A. K. A new approach for optimization of electrospun nanofiber formation process. *Korean J Chem Eng* **2010,** *27,* 340–354.
25. Shimada, I.; Takahagi, T. FT-IR Study of the Stabilization Reaction of Polyacrylonitrile in the Production of Carbon Fibers. *J. Polymer Science, Part A: Polymer Chemistry* **1986,** *24,* 1989–1995.
26. Ouyang, Q.; Cheng, L.; Wang, H.; Li, K. Mechanism and kinetics of the stabilization reactions of itaconic acid-modified polyacrylonitrile. *Polymer Degradation and Stability* **2008,** *93,* 1415–1421.
27. Kaburagi, M.; Bin Y, Zhu, D.; Xu, C,; Matsuo, M. Small angle Xray scattering from voids within fibers during the stabilization and carbonization stages. *Carbon* **2003,** *41,* 915–926.
28. Yu. M.; Wang, C.; Bai Y, Wang, Y.; Wang, Q.; Liu H. Combined Effect of Processing Parameters on Thermal Stabilization of PAN Fibers. *Polymer Bulletin* **2006,** *57,* 525–533.
29. Su, Y.; Ko T.; Lin, J. Preparation of Ultra-thin PAN-based Activated Carbon Fibers with Physical Activation. *J Appl Polym Sci* **2008,** *108,* 3610–3617.

**CHAPTER 6**

# CARBON NANOTUBES STRUCTURE IN POLYMER NANOCOMPOSITES

Z. M. ZHIRIKOVA, V. Z. ALOEV, G. V. KOZLOV, and G. E. ZAIKOV

## CONTENTS

## 6.1   INTRODUCTION

In this chapter, carbon nanotubes (Appendix) structure in polymer nano-composites was studied. It has been shown that this nanofiller feature is its "rolling up" in ring-like structures. This factor plays a crucial role in determination of nanocomposites structural and mechanical characteristics.

At present it is considered that carbon nanotubes (CNT) are one of the most perspective nanofillers for polymer nanocomposites [1]. The high anisotropy degree (their length to diameter large ratio) and low transverse stiffness are $CNT_s$ specific features. These factors define $CNT_s$ ring-like structures formation at manufacture and their introduction in polymer matrix. Such structures radius depends to a considerable extent on $CNT_s$ length and diameter. Thus, the strong dependence of nanofiller structure on its geometry is $CNT_s$ application specific feature. Therefore, the present work purpose is to study the dependence of nanocomposites butadiene-stirene rubber/carbon nanorubes (BSR/CNT) properties on nanofiller structure, received by CVD method with two catalysts usage.

## 6.2   EXPERIMENTAL METHODS

The nanocomposites BSR/CNT with CNT content of 0.3 mass% have been used as the study object. CNT have been received in the Institute of Applied Mechanics of Russian Academy of Sciences by the vapors catalytic chemical deposition method (CVD), based on carbon – containing gas thermochemical deposition on nonmetallic catalyst surface. Two catalysts – $Fe/Al_2O_3$ (CNT-fe) and $Co/Al_2O_3$ (CNT-co) – have been used for the studied CNT. The received nanotubes have diameter of 20 nm and length of order of 2 mcm.

The nanofiller structure was studied on force-atomic microscope Nano-DST (Pacific Nanotechnology, USA) by a semicontact method in the force modulation regime. The received CNT size and polydispersity analysis was made with the aid of the analytical disk centrifuge (CPS Instrument, Inc., USA), allowing to determine with high precision the size and distribution by sizes in range from 2 nm up to 5 mcm. The nanocomposites BSR/CNT elasticity modulus was determined by nanoindentation method on apparatus Nano-test 600 (Great Britain).

## 6.3   RESULTS AND DISCUSSION

In Fig. 1 the electron microphotographs of CNT coils are adduced, which demonstrate ring-like structures formation for this nanofiller. In Fig. 2, the indicated structures distribution by sizes was shown, from which it follows, that for CNT-Fe narrow enough monodisperse distribution with maximum at 280 nm is observed and for CNT-Co – polydisperse distribution with maximums at ~50 and 210 nm.

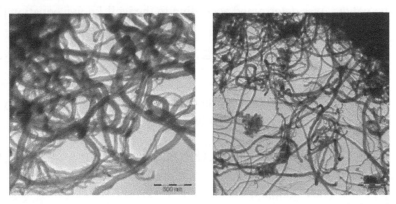

**FIGURE 1**    Electron micrographs of CNT structure, received on transmission electron microscope.

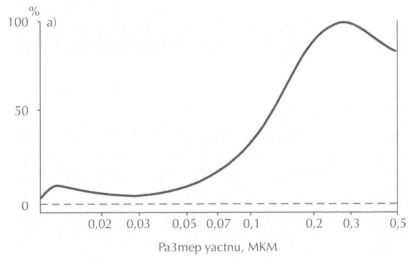

**FIGURE 2**    The particles distridution by sizes for CNT-Fe (1) and CNT-Co (2).

Further let us carry out the analytical estimation of CNT formed ring-like structures radius $R_{CNT}$. The first method uses the following formula, obtained within the frameworks of percolation theory [2]:

$$\phi_n = \frac{\pi L_{CNT} r_{CNT}^2}{\left(2R_{CNT}\right)^3} \tag{1}$$

where $j_n$ is CNT volume content, $L_{CNT}$ and $r_{CNT}$ are CNT length and radius, respectively.

The value $j_n$ was determined according to the well-known equation [3]:

$$\phi_n = \frac{W_n}{\rho_n} \tag{2}$$

where $fi_n$ and $r_n$ are mass content and density of nanofiller, respectively.

In its turn, the value $r_n$ was calculated as follows [4]:

$$\rho_n = 0.188 D_{CNT}^{1/3} \tag{3}$$

where $D_{CNT}$ is CNT diameter.

The second method is based on the following empirical formula application [5]:

$$R_{CNT} = \left(\frac{D_{CNT}}{D_{CNT}^{st}}\right)^2 \left(0.64 + 4.5 \times 10^{-3} \phi_n^{-1}\right), \text{mcm} \tag{4}$$

where $D_{CNT}^{st}$ is a standard nanotube diameter, accepted in paper [5] equal to CNT of the mark "Taunite" diameter (45 nm).

The values $R_{CNT}$, calculated according to the Eqs. (1) and (4), are adduced in Table 1, from which their good correspondence (the discrepancy is equal to ~15%) follows. Besides, they correspond well enough to Fig. 2 data.

**TABLE 1**    The structural and mechanical characteristics of nanocomposites BSR/CNT.

| Catalyst | $E_n$, MPa | $E_n/E_m$ | $E_n/E_m$, the Eq. (5) | $R_{CNT}$, nm, the Eq. (1) | $R_{CNT}$, nm, the Eq. (4) | $b$, the Eq. (6) |
|---|---|---|---|---|---|---|
| Fe/Al$_2$O$_3$ | 4.9 | 1.485 | 1.488 | 236 | 278 | 8.42 |
| Co/Al$_2$O$_3$ | 3.1 | ~1.0 | 1.002 | 236 | 278 | 0.27 |

**Footnote:** the value $E_m$ for BSR is equal to 3.3 MPa.

In Table 1, the values of elasticity modulus $E_n$ for the studied nano-composites and $E_m$ for the initial BSR are also adduced. As one can see, if for the nanocomposite BSR/CNT-Fe the very high (with accounting of the condition $fi_n$=0.3 mass%) reinforcement degree $E_n/E_m$=1.485 was obtained, then for the nanocomposite BSR/CNT-Co reinforcement is practically absent (with accounting for experiment error): $E_n \gg E_m$. Let us consider the reasons of such essential distinction.

As it is known [4], the reinforcement degree for nanocomposites polymer/CNT can be calculated as follows:

$$\frac{E_n}{E_m} = 1 + 11 \left( c\phi_n b \right)^{1.7}$$ (5)

where $c$ is proportionality coefficient between nanofiller $j_n$ and interfacial regions $j_{if}$ relative fractions, $b$ is the parameter, characterizing interfacial adhesion polymer matrix-nanofiller level.

The parameter $b$ in the nanocomposites polymer/CNT case depends on nanofiller geometry as follows [5]:

$$b = 80 \left( \frac{R_{CNT}^2}{L_{CNT} D_{CNT}^2} \right)$$ (6)

Calculated according to the Eq. (6) values $b$ (for BSR/CNT-Fe and BSR/CNT-Co $R_{CNT}$ magnitudes were accepted equal to 280 and 50 nm, respectively) are adduced in Table 1. As one can see, $R_{CNT}$ decreasing for the second from the indicated nanocomposites results to $b$ reduction in more than 30 times.

The coefficient $c$ value in the Eq. (5) can be calculated as follows [4]. First the interfacial layer thickness $l_{if}$ is determined according to the equation [6]:

$$l_{if} = a\left(\frac{r_{CNT}}{a}\right)^{2(d-d_{surf})/d} \qquad (7)$$

where $a$ is lower linear scale of polymer matrix fractal behavior, accepted equal to statistical segment length $l_{st}$, $d$ is dimension of Euclidean space, in which fractal is considered (it is obvious that in our case $d=3$), $d_{surf}$ is CNT surface dimension, which for the studied CNT was determined experimentally and equal to 2.89.

The indicated dimension $d_{surf}$ has a very large absolute magnitude ($2 \le d_{surf} < 3$ [4]) that supposes corresponding roughness of CNT surface, which BSR macromolecule, simulated by rigid statistical segments sequence [7], cannot be "reproduced." Therefore, in practice the effective value $d_{surf}$ ($d_{surf}^{ef}$ for $d_{surf} > 2.5$) is used, which is equal to [7]:

$$d_{surf}^{ef} = 5 - d_{surf} \qquad (8)$$

And at last, the statistical segment length $l_{st}$ is estimated according to the equation [4]:

$$l_{st} = l_0 C_\infty \qquad (9)$$

where $l_0$ is the length of the main chain skeletal bond, $C_\infty$ is characteristic ratio. For BSR $l_0=0.154$ nm, $C_\infty=12.8$ [8].

Further, simulating an interfacial layer as cylindrical one with external radius $r_{CNT}+l_{if}$ and internal radius $r_{CNT}$, let us obtain from geometrical considerations the formula for $j_{if}$ calculation [6]:

$$\phi_{if} = \phi_n\left[\left(\frac{r_{CNT}+l_{if}}{r_{CNT}}\right)^3 - 1\right] \qquad (10)$$

according to which the value $c$ is equal to 3.47.

The reinforcement degree $E_n/E_m$ calculation results according to the Eq. (5) are adduced in Table 1. As one can see, these results are very close to the indicated parameter experimental estimations. From the Eq. (5) it follows unequivocally, that the values $E_n/E_m$ distinction for nanocomposites BSR/CNT-Fe and BSR/CNT-Co is defined by the interfacial adhesion level difference only, characterized by the parameter $b$, since the values $c$ and $j_n$ are the same for the indicated nanocomposites. In its turn, from the Eq. (6) it follows so unequivocally, that the parameters $b$ distinction for the indicated nanocomposites is defined by $R_{CNT}$ difference only, since the values $L_{CNT}$ and $D_{CNT}$ for them are the same. Thus, the fulfilled analysis supposes CNT geometry crucial role in nanocomposites polymer/CNT mechanical properties determination.

Let us note, that the usage of the average value $R_{CNT}$ for nanocomposites BSR/CNT-Co according to Fig. 2 data in the Eq. (6), which is equal to 130 nm, will not change the conclusions made above. In this case $E_n/E_m=1.036$, that is again close to the obtained experimentally practical reinforcement absence for the indicated nanocomposite.

## 6.4 CONCLUSIONS

Thus, the obtained in the present work results have shown that the nanotubes geometry, characterized by their length, diameter and ring-like structures radius, is nanocomposites polymer/CNT specific feature. This factor plays a crucial role in the interfacial adhesion polymer matrix – nanofiller level determination and, as consequence, in polymer nanocomposites, filled with CNT, mechanical properties formation.

## APPENDIX

### A.1 INTRODUCTION

In 1991, Japanese researchers studied sediment formed at the cathode during the spray of graphite in an electric arc. Their attention was attracted by the unusual structure of the sediment consisting of microscopic fibers and filaments. Measurements made with an electron microscope showed that the diameter of these filaments do not exceed a few nanometers and a length of from one to several microns.

Having managed to cut a thin tube along the longitudinal axis, the researchers found that it consists of one or more layers, each of which represents a hexagonal grid of graphite, which is based on hexagon with vertices located at the corners of the carbon atoms. In all cases, the distance between the layers is equal to 0.34 nm, which is the same as that between the layers in crystalline graphite.

Typically, the upper ends of tubes are closed by multilayer hemispherical caps; each layer is composed of hexagons and pentagons, reminiscent of the structure of half a fullerene molecule.

The extended structure consisting of rolled hexagonal grids with carbon atoms at the nodes, called nanotubes.

Lattice structure of diamond and graphite are shown in Fig. A.1. Graphite crystals are built of planes parallel to each other, in which carbon atoms are arranged at the corners of regular hexagons. The distance between adjacent carbon atoms (each side of the hexagon) $d_0 = 0{,}141m$ , between adjacent planes 0.335 nm.

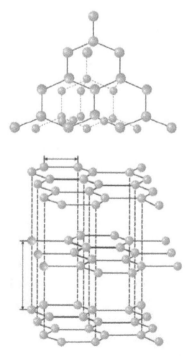

**FIGURE A.1**    The structure of (a) the diamond lattice, and (b) graphite.

Each intermediate plane is shifted somewhat toward the neighboring planes, as shown in the Fig. A.1.

The elementary cell of the diamond crystal is a tetrahedron in the center and four vertices of which are carbon atoms. Atoms located at the vertices of a tetrahedron, form a new center of a tetrahedron, and thus, also surrounded by four atoms each, etc. All the carbon atoms in the crystal lattice are located at equal distance (0.154 nm) from each other.

Nanotubes rolled into a cylinder (hollow tube) graphite plane, which is lined with regular hexagons with carbon atoms at the vertices of a diameter of several nanometers (Fig. A.2). Nanotubes can consist of one layer of atoms (single-wall nanotubes-SWNT) and represent a number of "nested" into one another-layer pipes (multiwalled nanotubes – MWNT).

Nanostructures can be collected not only from individual atoms or single molecules, but the molecular blocks. Such blocks or elements to create nanostructures are graphene, carbon nanotubes and fullerenes.

## A.2  GRAPHENE

Graphene is a single flat sheet, consisting of carbon atoms linked together and forming a grid, each cell is like a bee's honeycombs (Fig. A.2). The distance between adjacent carbon atoms in graphene is about 0.14 nm.

**FIGURE A.2**    Schematic illustration of the grapheme.

Graphite, from which are made slates of usual pencils, is a pile of graphene sheets (Fig. A.3). Graphenes in graphite is very poorly connected and can slide relative to each other. So, if you conduct the graphite on paper, then after separating graphene from sheet the graphite remains on paper. This explains why graphite can write.

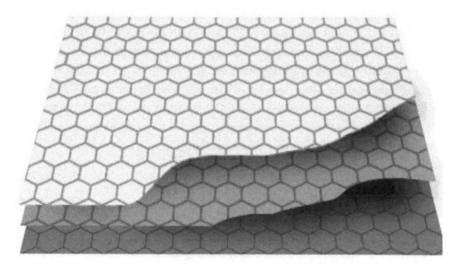

**FIGURE A.3**   Schematic illustrations of the three sheets of grapheme.

## A.3   CARBON NANOTUBES

Many perspective directions in nanotechnology are associated with carbon nanotubes.

*Carbon nanotubes*: a carcass structure or a giant molecule consisting only from carbon atoms (Fig. A.4).

Carbon nanotube is easy to imagine, if we imagine that you fold up one of the molecular layers of graphite – graphene (Fig. A.5).

**FIGURE A.4**    Carbon nanotubes.

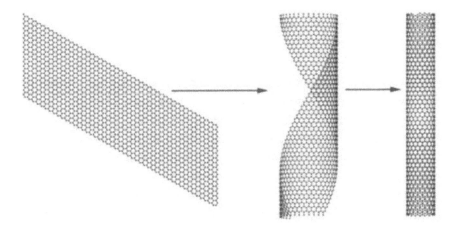

**FIGURE A.5**    Imaginary making nanotube (right) from the molecular layer of graphite (left).

Nanotubes formed themselves, for example, on the surface of carbon electrodes during arc discharge between them. At discharge, the carbon atoms evaporate from the surface and connecting with each other to form nanotubes of all kinds; single, multilayered and with different angles of twist (Fig. A.6).

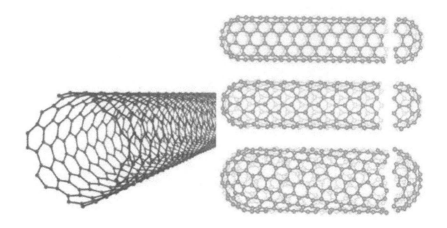

**FIGURE A.6**  Schematic representation of a single-layer carbon nanotubes, on the right (top to bottom) – two-ply, straight and spiral nanotubes.

The diameter of nanotubes is usually about 1 nm and their length is a thousand times more, amounting to about 40 microns. They grow on the cathode in perpendicular direction to surface of the butt. Occur so is called self-assembly of carbon nanotubes from carbon atoms. Depending on the angle of folding of the nanotube can have a high as that of metals, conductivity, and can have properties of semiconductors.

Carbon nanotubes are stronger than graphite, although made of the same carbon atoms, because the carbon atoms in graphite are in the sheets. And everyone knows that folding into a tube sheet of paper is much more difficult to bend and break than a regular sheet. That's why carbon nanotubes are strong. Nanotubes can be used as a very strong microscopic rods and filaments, as Young's modulus of single-walled nanotube reaches values of the order of 1–5 TPa, which is much more than steel! Therefore, the thread made of nanotubes, the thickness of a human hair is capable to hold down hundreds of kilos of cargo.

It is true that at present the maximum length of nanotubes is usually about a 100 microns, which is certainly too small for everyday use. However, the length of the nanotubes obtained in the laboratory, gradually increasing – now scientists have come close to the millimeter border. So there is every reason to hope that in the near future, scientists will learn how to grow a nanotube length in centimeters and even meters!

## A.4 FULLERENES

The carbon atoms, evaporated from a heated graphite surface, connecting with each other, can form not only of the nanotube, but also other molecules, which are closed convex polyhedra, for example, in the form of a sphere or ellipsoid. In these molecules, the carbon atoms located at the vertices of regular hexagons and pentagons that make up the surface of a sphere or ellipsoid.

All of these molecular compounds of carbon atoms called fullerenes on behalf of the American engineer, designer and architect R. Buckminster Fuller's domes were used for construction of its buildings, pentagons and hexagons (Fig. A.7), which are the main structural elements of the molecular carcasses of all of fullerenes.

**FIGURE A.7**  Biosphere of Fuller (Montreal, Canada).

The molecules of the symmetrical and the most studied fullerene consisting of 60 carbon atoms ($C_{60}$), form a polyhedron consisting of 20 hexagons and 12 pentagons and resembles a soccer ball (Fig. A.8). The diameter of the fullerene $C_{60}$ is about 1 nm.

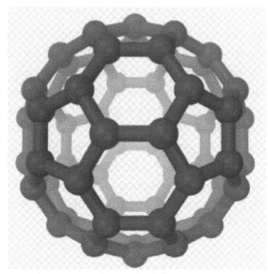

**FIGURE A.8**    Schematic representation of the fullerene $C_{60}$.

## A.5   CLASSIFICATION OF NANOTUBES

The main classification of nanotubes is conducted by the number of constituent layers.

**Single-walled nanotubes**: The simplest form of nanotubes (Fig. A.9). Most of them have a diameter of about 1 nm in length, which can be many thousands of times more. The structure of the nanotubes can be represented as a "wrap" a hexagonal network of graphite (graphene), which is based on hexagon with vertices located at the corners of the carbon atoms in a seamless cylinder. The upper ends of the tubes are closed by hemispherical caps, each layer is composed of six- and pentagons, reminiscent of the structure of half of a fullerene molecule. The distance d between adjacent carbon atoms in the nanotube is approximately equal to $d = 0.15$ nm.

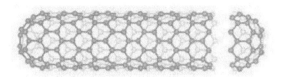

**FIGURE A.9**   Graphical representation of single-walled nanotube.

**Multi-walled nanotubes** consist of several layers of graphene stacked in the shape of the tube (Fig. A.10). The distance between the layers is equal to 0.34 nm, that is the same as that between the layers in crystalline graphite.

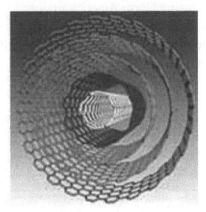

**FIGURE 1.10**   Graphic representation of a multiwalled nanotube.

Due to its unique properties (high fastness (63 GPa), superconductivity, capillary, optical, magnetic properties, etc.), carbon nanotubes could find applications in numerous areas:
- Additives in polymers;
- Catalysts (autoelectronic emission for cathode ray lighting elements, planar panel of displays, gas discharge tubes in telecom networks);
- Absorption and screening electromagnetic waves;
- Transformation of energy;
- Anodes in lithium batteries;
- Keeping of hydrogen;
- Composites (filler or coating);

- Nanosondes;
- Sensors;
- strengthening of composites;
- Supercapacitors.

More than a decade, carbon nanotubes, despite their impressive performance characteristics have been used, in most cases, for scientific research. These materials are not yet able to gain a foothold in the market, mainly because of problems with their large-scale production and uncompetitive prices.

To date, the most developed production of nanotubes has Asia, the production capacity, which is 2–3 times higher than in North America and Europe combined. Is dominated by Japan, which is a leader in the production of MWNT. Manufacturing North America mainly focused on the SWNT. In the coming years, China will surpass the level of production of the U.S. and Japan, and by now, a major supplier of all types of nanotubes, according to experts, could be South Korea.

## A.6   CHIRALITY

*Chirality is* a set of two integer positive indices $(n, m)$, which determines how the folds the graphite plane and how many elementary cells of graphite at the same time fold to obtain the nanotube.

From the value of parameters $(n, m)$ are distinguished
- direct (achiral) high-symmetry carbon nanotubes
  - o   armchair $n = m$
  - o   zigzag $m = 0$ or $n = 0$
  - o   helical (chiral) nanotube

Figure A.11a shows a schematic image of the atomic structure of graphite plane – graphene, and shows how from it can be obtained the nanotube. The nanotube is fold up with the vector connecting two atoms on a graphite sheet. The cylinder is obtained by folding this sheet so that were combined the beginning and end of the vector. That is, to obtain a carbon nanotube from a graphene sheet, it should turn so that the lattice vector $\overline{R}$ has a circumference of the nanotube in Fig. A.11b. This vector can be expressed in terms of the basis vectors of the elementary cell graphene sheet $\vec{R} = n\vec{r_1} + m\vec{r_2}$ . Vector $\overline{R}$, which is often referred to simply by a pair of indices $(n, m)$, called the chiral vector. It is assumed that $n > m$. Each pair of numbers $(n, m)$ represents the possible structure of the nanotube.

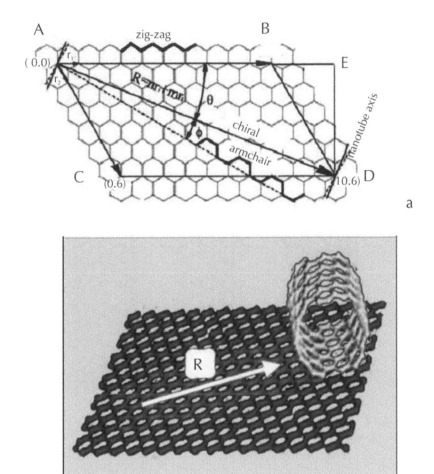

**FIGURE A.11**  Atomic structure of graphite plane.

In other words the chirality of the nanotubes $(n, m)$ indicates the coordinates of the hexagon, which as a result of folding the plane has to be coincide with a hexagon, located at the beginning of coordinates (Fig. A.12).

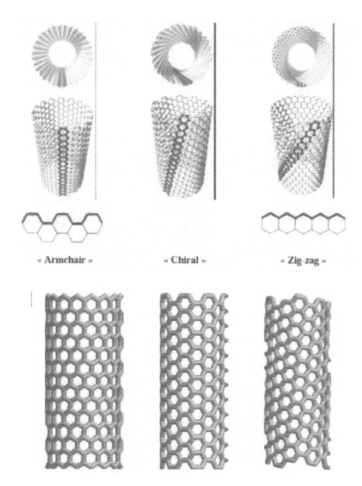

**FIGURE A.12**   Single-walled carbon nanotubes in different chirality, Left to right: the zigzag (16, 0), armchair (8, 8) and chiral (10, 6) carbon nanotubes.

Many of the properties of nanotubes (e.g., zonal structure or space group of symmetry) strongly depend on the value of the chiral vector. Chirality indicates what property has a nanotube – a semiconductor or metallicheskm. For example, a nanotube (10, 10) in the elementary cell contains 40 atoms and is the type of metal, whereas the nanotube (10, 9) has already in 1084 and is a semiconductor (Fig. A.13).

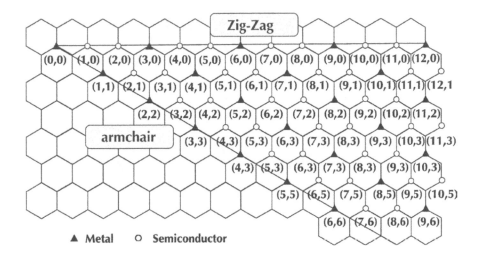

**FIGURE A.13** The scheme of indices $(n, m)$ of lattice vector $\overline{R}$ tubes having semiconductor and metallic properties.

If the difference $n - m$ is divisible by 3, then these CNTs have metallic properties. Semimetals are all achiral tubes such as "chair". In other cases, the CNTs show semiconducting properties. Just type chair CNTs $(n = m)$ are strictly metal.

## A.7 DIAMETER, CHIRALITY ANGLE AND THE MASS OF SINGLE-WALLED NANOTUBE

Indices $(n, m)$ of single-walled nanotube chirality unambiguously determine its diameter. Therefore, the nanotubes are typically characterized by a diameter and chirality angle. Chiral angle of nanotubes is the angle between the axis of the tube and the most densely packed rows of atoms. From geometrical considerations, it is easy to deduce relations for the chiral angle and diameter of the nanotube. The angle between the basis vectors of the elementary cell (Fig. A.14) $\vec{r_1}$ and $\vec{r_2}$ is equal 60°.

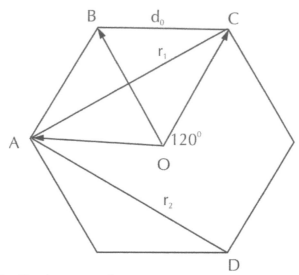

**FIGURE A.14**   The elementary cell.

As we know from trigonometry, $AC^2 = OA^2 + OC^2 - 2OA \cdot OC \cdot \cos 120^0$. As $OA = OC = d_0$, a $r_1 = r_2 = AC$, we have

$$r_1 = r_2 = \sqrt{3} \cdot d_0 \qquad (A.1)$$

where $d_0 = 1{,}41 \overset{0}{A} = 0{,}141 \textit{нм} -$ distance between neighboring carbon atoms in the graphite plane. Thus, the basis vectors $\vec{r_1}$ и $\vec{r_2}$ of the elementary cell of graphene are $\left|\vec{r_1}\right| = \left|\vec{r_2}\right| = 0{,}244$.

Now consider the parallelogram $ABDC$ in Fig. A.11a.

According to Eq. (A.1) we have

$$AB = CD = \sqrt{3}d_0 n \quad AC = BD = \sqrt{3}d_0 m \qquad (A.2)$$

Angle $\angle CAB = 60^0$, and $\angle ABD = 120^0$, therefore

$R^2 = 3n^2 d_0^2 + 3m^2 d_0^2 - 2 \cdot 3mnd_0^2 \cos 120^0$, from which we obtain

$$R = \sqrt{3}d_0 \sqrt{n^2 + m^2 + mn}$$

Taking into account that $R = \pi \cdot d$, then to determine the diameter of the nanotube we obtain the expression

$$d = \frac{|\vec{R}|}{\pi} = \sqrt{3(m^2 + n^2 + mn)} \cdot \frac{d_0}{\pi} \qquad (A.3)$$

When $m = n$ we have

$$d = \frac{3nd_0}{\pi}$$

Below in Table A.1 shows the values of the diameters of nanotubes of different chirality.

Thus, knowing $d$ the chirality can be found $m$ and $n$ (possible relations $m$ and $n$, Table A.2). The minimum diameter of the tube is close to 0.4 nm, which corresponds to the chirality (3, 3), (5, 0), (4, 2). Unfortunately, the objects of that the diameter of the least stable. Of single-walled nanotube was most stable with chirality indices (10, 10), its diameter is equal 1.35 nm.

**TABLE A.1**  Diameters of nanotubes of different chirality.

| (n, m) | d, nm | (n, m) | d, nm |
|--------|-------|--------|-------|
| (3,2)  | 0,334 | (10,8) | 1,232 |
| (4,2)  | 0,417 | (10,9) | 1,298 |
| (4,3)  | 0,480 | (11,3) | 1,007 |
| (5,0)  | 0,394 | (11,6) | 1,177 |
| (5,1)  | 0,439 | (11,10)| 1,434 |
| (5,3)  | 0,552 | (12,8) | 1,375 |
| (6,1)  | 0,517 | (14,13)| 1,844 |
| (7,3)  | 0,701 | (20,19)| 2,663 |
| (9,2)  | 0,801 | (21,19)| 2,732 |
| (9,8)  | 1,161 | (40,38)| 5,326 |

**TABLE A.2**  CNT with of different chirality.

| CNT (n, m) | Diameter CNT, nm | Chirality |
|---|---|---|
| (4,0) | 0,33 | |
| (5,0) | 0,39 | |
| (6,0) | 0,47 | |
| (7,0) | 0,55 | |
| (8,0) | 0,63 | zigzag |
| (9,0) | 0,70 | |
| (10,0) | 0,78 | |
| (11,0) | 0,86 | |
| (12,0) | 0,93 | |
| (3,3) | 0,40 | |
| (4,4) | 0,56 | |
| (5,5) | 0,69 | armchair |
| (6,6) | 0,81 | |
| (7,7) | 0,96 | |
| (8,8) | 1,10 | |
| (4,1) | 0,39 | |
| (4,2) | 0,43 | |
| (7,1) | 0,57 | |
| (6,3) | 0,62 | chiral |
| (9,1) | 0,75 | |
| (10,1) | 0,82 | |
| (6,7) | 0,90 | |

We derive a formula for determining the mass of the nanotube with diameter $d$, length $L$.

The area of the elementary area – a parallelogram with vertices at the centers of four neighboring hexagons (Fig. A.15) with base $\sqrt{3}d_0$ and height $3d_0/2$ is equal $S_{n\pi} = \dfrac{3\sqrt{3}}{2}d_0^2$.

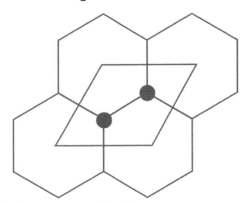

**FIGURE A.15**   The elementary area of grapheme.

The total area of the nanotube is $\pi dL$. Consequently, the number of elementary areas is equal $\pi dL / S_{n\pi}$. At the same time in each elementary site contains two carbon atoms. Consequently, the number of carbon atoms in the tube is twice more than the number of elementary areas that can fit on the surface. Therefore, the mass of a carbon nanotube is equal to:

$$m_T = 2m_C \frac{\pi L d}{S_{n\pi}} = \frac{4\sqrt{3}\pi \cdot dL}{9d_0^2} m_C \tag{A.4}$$

where $m_C = 12$ – mass of carbon atoms.

To determine the chiral angle $\theta$ from a right triangle $AED$ we obtain

$$\sin\theta = \frac{DE}{R}, \quad \cos\theta = \frac{AE}{R} = \frac{\sqrt{3}nd_0 + BE}{R}$$

If we take into consideration that $\angle EDB = 30^0$, we see that $DE = \dfrac{3}{2}md_0$ and $BE = \dfrac{\sqrt{3}}{2}md_0$, consequently,

$$\sin\theta = \frac{3md_0}{2R}, \quad \cos\theta = \frac{\sqrt{3}d_0(n + m/2)}{R}$$

From these equalities we obtain the relation between the chiral indices $(m,n)$ and angle $\theta$:

$$\theta = arctg\left(\frac{\sqrt{3}m}{2n+m}\right) \tag{A.5}$$

When $m = n$ we have

$$\theta = arctg\frac{\sqrt{3}}{3}$$

## KEYWORDS

- **carbon nanotubes**
- **chiral indices**
- **Euclidean space**
- **polymer matrix**
- **ring-like structures**

## REFERENCES

1. Yanovskii, Yu.G. Nanomechanics and Strength of Composite Materials. Moscow, Publishers of IPRIM RAN, **2008**, 179.
2. Bridge, B. *J. Mater. Sci. Lett.*; **1989,** *8(2),* 102–103.
3. Sheng N.; Boyce, M. C.; Parks, D. M.; Rutledge, G. C.; Abes, J. I.; Cohen R. E.; *Polymer*, **2004,** *45(2),* 487–506.
4. Mikitaev, A. K.; Kozlov, G. V.; Zaikov, G. E. Polymer Nanocomposites: Variety of Structural Forms and Applications. Nova Science Publishers, Inc.: New York, **2008,** 319.
5. Zhirikova, Z. M.; Kozlov, G. V.; Aloev Mater, V. Z.; of VII Intern. Sci.-Pract. Conf. "New Polymer Composite Materials." Nal'chik, KBSU, **2011,** 158–164.
6. Kozlov, G. V.; Burya, A. I.; Lipatov Yu.S. Mekhanika Kompozitnykh Materialov, **2006,** *42(6),* 797–802.
7. Van Damme, H.; Levitz; Bergaya, F.; Alcover, J. F.; Gatineau L.; Fripiat J. J.; J. Chem. Phys.; **1986,** *85(1),* 616–625.
8. Yanovskii Yu.G.; Kozlov, G. V.; Karnet, Yu.N. Mekhanika Kompozitsionnykh Materialov i Konstruktsii, **2011,** *17(2),* 203–208.

**CHAPTER 7**

# EXPLORING THE POTENTIAL OF OILSEEDS AS A SUSTAINABLE SOURCE OF OIL AND PROTEIN FOR AQUACULTURE FEED

CRYSTAL L. SNYDER, PAUL P. KOLODZIEJCZYK, XIAO QIU, SALEH SHAH, E. CHRIS KAZALA, and RANDALL J. WESELAKE

## CONTENTS

## 7.1　INTRODUCTION

Aquaculture is currently the world's fastest growing primary food production sector, but it is also a leading consumer of fish oil and fish meal from capture fisheries, which is a major concern for the industry's long-term sustainability. Oilseeds have been widely explored as an alternative, land-based source of oil and protein for aquaculture feeds, but their use has been limited because conventional seed oils do not contain very long chain $\omega$-3 fatty acids such as eicosapentaenoic acid (EPA) and docosahexaenoic acid (DHA). Over the past decade, however, there has been considerable progress toward the genetic engineering of oilseeds to produce very long chain $\omega$-3 fatty acids. Such modified oilseeds show promise as nutritional supplements for aquafeed applications. Advances in oilseed processing technology may also assist in the development of sustainable oilseed-based aquafeeds.

Aquaculture currently supplies almost half of the seafood for human and animal consumption [1], and as one of the world's fastest growing food production sectors, the industry is under pressure to address the sustainability of its supply chain, particularly with respect to its reliance on fish oil and fish meal from capture fisheries [2]. The growing concern over the decline of the world's capture fisheries has prompted considerable research into alternative aquafeed ingredients, especially those from land-based plants. Oilseeds, for example, have been widely explored as an alternative source of oil and protein in fish diets, but present a number of challenges that currently limit their use in aquafeeds. Unlike marine oils, which are enriched in very long chain (20–22 carbon) polyunsaturated fatty acids (VLCPUFA) such as eicosapentaenoic acid (EPA, 20:5$\omega$-3) and docosahexaenoic acid (DHA, 22:6$\omega$-3), conventional seed oils are rich in 18-carbon fatty acids (mainly linoleic acid, 18:2$\omega$-6 and $\alpha$-linolenic acid, 18:3$\omega$-3), which fish are unable to efficiently convert to VLCPUFA. In addition, animals lack the ability to convert omega-6 fatty acids to omega-3 fatty acids, making most high linoleic vegetable oils an inadequate substitute for fish oil. Maintaining high levels of $\omega$-3 VLCPUFA in farmed fish is of paramount concern not only for optimal fish health, but also for the nutritional quality of the end products destined for human consumption. Indeed, the relatively low consumption of $\omega$-3 VLCPUFA in the Western diet, together with high intake of $\omega$-6 fatty acids, has been blamed in part for the growth of chronic, diet-related disease among Western populations [3].

While oilseeds offer many advantages for the sustainable production of aquafeed ingredients, engineering oilseeds to produce VLCPUFA has proven to be a formidable biotechnological challenge. Over the past decade, research groups around the world have explored various strategies for producing some or all of the fatty acids in the VLCPUFA biosynthetic pathway [4]. Almost all of these approaches require the insertion and coordinated expression of multiple trans-genes, the effectiveness of which tends to vary depending on the nature of the genetic construct and of the host plant [5, 6]. To date, substantial accumulations of various precursors leading to the formation of arachidonic acid (ARA, 20:4$\omega$-6) and EPA have been obtained [4]; however, commercially viable levels of DHA have yet to be achieved.

Although much of the commercial value of commodity oilseeds is in the oil fraction, the meal of many oilseeds, if processed appropriately, can be used as a high-quality protein source for animal feeds. Research is ongoing into the use and enhancement of oilseed meal for aquafeed, with many studies focusing on the enrichment of high-value bioactive components (e.g., carotenoids) or reduction of antinutritional factors [7]. At the same time, advances in oilseed processing technology are also helping to maximize the value of the oil and the meal for aquafeed and other applications.

This paper will review progress toward the production of $\omega$-3 fatty acid-enriched seed oils for aquaculture feeds, their efficacy as aquafeed ingredients, and the role of processing technologies in maximizing the potential of oilseeds for aquaculture applications.

## 7.2  ENGINEERING OILSEEDS FOR THE SUSTAINABLE PRODUCTION OF $\Omega$-3 FATTY ACIDS FOR AQUACULTURE

In animals, linoleic acid (LA, 18:2 $\omega$-6) and $\alpha$-linolenic acid (ALA, 18:3 $\omega$-3) are considered to be the only dietary fatty acids that are essential for the production of VLCPUFA, including EPA and DHA. Animals possess a biosynthetic pathway for the further desaturation and elongation of linoleic and $\alpha$-linolenic acids to form ARA and EPA, respectively (Fig. 1A). However, the poor efficiency of this process, particularly the initial $\Delta 6$

desaturation step, can severely limit the accumulation of VLCPUFA in many individuals [8]. As a result, direct consumption of dietary VLCPU-FA is recommended for optimal nutrition [9]. Even fish, which are generally considered to be excellent dietary sources of VLCPUFA for humans, must obtain their VLCPUFA from their diet, usually through bioaccumulation of VLCPUFA produced by marine algae and other microorganisms.

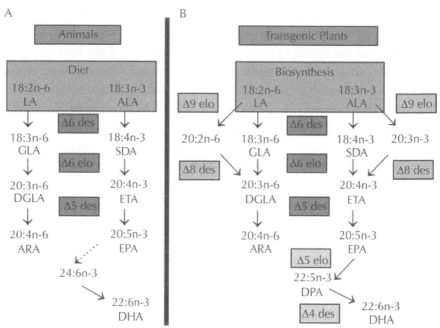

**FIGURE 1** Pathways for very long chain polyunsaturated fatty acid (VLCPUFA) biosynthesis in animals and plants. (A) In vertebrates, VLCPUFA are synthesized from the dietary essential fatty acids, linoleic acid (LA) and α-linolenic acid (ALA), by the Δ6-desaturation pathway, consisting of a Δ6-desaturase Δ6-elongase, and a Δ5-desaturase, leading to the formation of arachidonic acid (ARA; ω-6 pathway) and eicosapentaenoic acid (EPA; ω-3 pathway). Docosahexaenoic acid (DHA) is synthesized from EPA following further elongation and desaturation to 24:6, which undergoes one round of β-oxidation to form DHA. In most vertebrates, however, the Δ6-desaturation step is limiting, resulting in poor rates of conversion to EPA and DHA. (B) Two desaturation/elongation pathways have been tested extensively in transgenic plants. The Δ6-desaturation pathway (red) is analogous to the pathway present in animals, but uses LA and ALA generated through *de novo* biosynthesis. The Δ9-elongation pathway (blue) also leads to the formation of ARA and EPA, but elongation takes place first, and is followed by two desaturation steps. From EPA, at least two additional activities (purple) are required to generate DHA. (des: desaturase; elo: elongase. All fatty acid abbreviations are defined in the text. The notation n-3/n-6 and ω-3/ω-6 are used interchangeably.)

The VLCPUFA-producing pathways of these marine microorganisms have offered a diverse array of genetic tools for oilseed biotechnology. Unlike animals, these microorganisms possess multiple pathways for VLCPUFA biosynthesis. Some organisms rely on aerobic desaturation/ elongation processes analogous to the pathway present in animals [10], while others use a polyketide synthase (PKS) to build VLCPUFA in a process similar to *de novo* fatty acid biosynthesis in plants [11]. A few higher plants, including borage and echium [12, 13], also possess the critical Δ6 desaturation activity and naturally accumulate gamma-linolenic acid (GLA, 18:3ω-6) and stearidonic acid (SDA, 18:4 ω-3); however, these plants lack the enzyme activities required to convert GLA or SDA to downstream VLCPUFA. Nevertheless, these plants have served as an important source of genes supporting oilseed biotechnology efforts.

Indeed, the Δ6 desaturation pathway has been one of the most extensively studied in transgenic plants. Expression of recombinant desaturases from various sources has resulted in relatively high accumulation of GLA, and to a lesser extent, SDA. In *Brassica juncea,* expression of a Δ6 desaturase alone yielded up to 40% GLA and 8% SDA [14]. Other groups have used additional Δ12 or Δ15 desaturases to increase the supply of LA or ALA precursors, respectively [15, 16]. Using a high linoleic safflower (*Carthamus tinctorius*) line coexpressing a Δ6 and Δ12 desaturase, Nykiforuk et al., [5] were able to obtain more than 70% GLA in the seed oil; these lines have recently been approved for commercialization as Sonova™ 400. Achieving high levels of SDA, particularly without significant coproduction of GLA, has been somewhat more challenging. To this end, an ω-3-selective Δ6 desaturase was introduced into flaxseed in order to preferentially increase the accumulation of SDA [17]. SDA-enriched seed oils, having bypassed the limiting Δ6 desaturation step, may be particularly useful for aquafeed applications; however, the efficacy of these oils is still being tested in fish [13, 18].

Conversion of GLA and SDA to ARA and EPA, respectively, requires the introduction of at least two additional genes encoding a Δ6 elongase and a Δ5 desaturase to complete the "conventional" Δ6 desaturation pathway (Fig. 1B). This pathway was first reconstituted in tobacco (high LA background) and flax (high ALA background), resulting in approximately 30% GLA or SDA, respectively, but only up to about 5% total 20-carbon

VLCPUFA [19]. It appeared that the $\Delta6$ elongation step was not very efficient, even though the recombinant elongase was successfully expressed in developing seeds and exhibited *in vitro* enzyme activity. Analysis of the acyl-CoA pool of transgenic flaxseed, however, indicated that there were only trace levels of $\Delta6$ desaturated PUFA in the acyl-CoA pool, where elongation occurs [19]. Since the $\Delta6$ desaturase uses phospholipid substrates and the elongase requires acyl-CoA substrates [20], this so-called "substrate dichotomy" was suggested to be a major limiting factor for VLCPUFA formation in transgenic plants [19]. These observations were supported by another study [21] employing an "alternative" $\Delta9$ elongation pathway (Fig. 1B), in which LA and ALA are first elongated, and then desaturated twice to form ARA and EPA. In this pathway, there is no need to transfer of substrates from phospholipids to the acyl-CoA pool and then back again. When this pathway was introduced into Arabidopsis leaves, up to 7% ARA and 3% EPA was obtained, and the total levels of 20-carbon PUFA were greater than 20% [21].

Other groups have tested a number of variations on the $\Delta6$ desaturation pathway in efforts to overcome this problem of substrate dichotomy. Adding a $\Delta12$ desaturase and an additional $\Delta6$ elongase to the minimal three gene pathway resulted in increased elongation efficiency in transgenic *Brassica juncea* seeds, yielding up to 25% ARA [22]. The further introduction of an $\omega$-3 desaturase, which converts $\omega$-6 intermediates to their $\omega$-3 counterparts, resulted in up to 11% EPA [22]. Zero erucic acid *Brassica carinata*, however, seems to be a more suitable host for EPA production, accumulating up to 20% EPA in lines expressing a five gene construct ($\Delta12$ desaturase, $\Delta6$ desaturase, $\Delta6$ elongase, $\Delta5$ desaturase, and an $\omega$-3 desaturase) [6]. In zero-erucic acid *B. juncea*, the same construct yielded only 5% EPA[6], which highlights the importance of host species on the outcome of such transgenic experiments.

Some groups have also explored the use of acyl-CoA desaturases as another approach to overcoming substrate dichotomy in transgenic plants. In this strategy, $\Delta6$ desaturation and elongation both occur in the acyl-CoA pool. Transgenic soybeans expressing an acyl-CoA/phospholipid desaturase from liverwort accumulated more than 19% total 20-carbon PUFA [23]. Although this increase comprised mostly $\omega$-6 fatty acids, reflecting the high levels of linoleic acid precursor in soybean, this study demonstrated the effectiveness of acyl-CoA desaturases in overcoming substrate dichotomy. Another study using an ω-3-specific acyl-CoA desaturase showed a similar proof-of-concept in *Arabidopsis*, but achieved much lower total levels of VLCPUFA [24].

While it appears that the nature of the genetic construct, source of the trans-genes, and selection of the host species all play a crucial role in determining the success of VLCPUFA-engineering experiments, recent advancements in our basic understanding of plant lipid biochemistry are also shedding insights into the biochemical basis for substrate dichotomy. The discovery of new enzymes involved in substrate trafficking [25, 26] and the elucidation of a proposed "acyl-editing" pathway in developing oilseeds [27, 28] will greatly assist ongoing efforts to develop a sustainable land-based source of VLCPUFA.

Although much of the research to date has focused on EPA as an essential first step in achieving high levels of VLCPUFA in transgenic plants, it is well recognized that DHA is equally important for aquaculture applications Several studies have shown proof-of-concept production of DHA in transgenic systems. Production of DHA has been technically more difficult, both in terms of overcoming substrate dichotomy and in handling the very large multigene constructs required for DHA production. Wu et al. [22], for example, used a nine-gene construct to demonstrate proof-of-concept DHA synthesis in *B. juncea*. Recent methodological advancements such as the trans-ient leaf-based assay described by Wood et al[29], promise to accelerate development of complex multigene constructs by facilitating rapid testing and interchangeability of independent genetic elements. This technique has already been successfully applied to the VLCPUFA pathway, resulting in up to 2.5% DHA in tobacco leaves and similar levels in stably transformed *Arabidopsis* seeds [30]. Given the rapid pace of research in this area, it is likely only a matter to time before commercially viable levels of DHA are achieved in a crop plant [31–34].

## 7.3   EFFICACY OF Ω-3 VEGETABLE OILS IN FISH NUTRITION

Over the past few decades, a wide range of vegetable oils and vegetable oil blends have been explored as possible alternatives to fish oil for aquafeed applications (*see* [35, 36] for recent reviews). All vertebrates, including fish, have certain dietary requirements for essential fatty acids (EFA). In fish, dietary EFA requirements differ according to species; freshwater and diadromous fish may only require 18-carbon PUFA (e.g., LA or ALA), while marine species require ω-3 VLCPUFA (e.g., EPA and DHA) [13]. A deficiency in EFA can lead to adverse effects on growth and reproduction and in severe cases, could result in increased mortality [37]. The diversity of vegetable oils available and the differences in the dietary needs of cultivated fish species have led to a variety of different strategies for the incorporation of vegetable oils into aquaculture feeds. Some of these studies will be reviewed here.

Although flaxseed is a relatively minor commodity oilseed in terms of total production, it has received a great deal of attention for aquafeed applications due to its high levels of ALA (50–60%). One of the earliest fish oil replacement studies compared flax oil, rendered animal fat, and salmon oil in the dry diet of Chinook salmon (*Onchorhynchous tshawytscha*) over 16 weeks [38]. Fish growth was not affected by the substitution, but the flesh fatty acid profiles, particularly PUFA content, varied according to the fatty acid composition of the diet. Several other fish oil replacement studies have reported similar trends; that is, the tissue fatty acid composition largely reflects that of the diet, but vegetable oils generally do not impair growth or performance of the fish [39–41]. After comparing two base diets (100% fish oil or 100% flax oil) substituted with various proportions of sunflower oil, Menoyo et al. [40], concluded that flax could be a viable source of dietary lipids for farmed Atlantic salmon. In trout (*Onchorhynchus mykiss*), studies comparing salmon oil, flax oil, soybean oil, poultry fat, pork lard and beef tallow suggested that the EPA and DHA levels in the tissue were regulated and maintained at 10–12% regardless of diet [41].

As a Δ6 desaturated ω-3 PUFA, stearidonic acid (SDA) is also particularly interesting, not only for aquaculture, but for human and livestock nutrition as well. In mammals, the conversion rate of SDA to EPA is around 17–33% [42–44], while the conversion of ALA to EPA is much lower at 0.2–7% [45]. The low rate of conversion of ALA to EPA is believed to be due to the inefficient Δ6 desaturation step; thus, consumption of SDA could effectively bypass this step and support more efficient accumulation of downstream VLCPUFA. As fish use a similar pathway for converting

ALA to EPA, SDA-enriched oils may be useful for maintaining or increasing EPA levels in the tissue.

To date, most studies investigating the efficacy of SDA have used echium oil as a source of SDA. Echium is one of a few higher plants that accumulate appreciable amounts of SDA (up to 14%). In Arctic charr (*Salvelinus alpines*), replacing 80% of the dietary fish oil with echium oil for 16 weeks led to increases in dihomo-gamma-linolenic acid (DGLA, 20:3ω-6) and eicosatetraenoic acid (ETA, 20:3ω-3), the immediate elongation products of GLA and SDA, respectively, in both flesh and liver tissue [46]. The echium diet, however, was not effective at maintaining EPA and DHA levels in charr. In a similar trial with Atlantic cod (*Gadus morhua*), a 100% echium oil diet for 18 weeks resulted in only slight increases in DGLA, with no increase in ETA [18]. Such studies illustrate the complexity of species-specific differences in dietary fatty acid metabolism.

In Atlantic salmon parr, neither echium oil nor a 1:1 blend of echium and canola oil resulted in a decrease in white muscle DHA content after 42 days [47]. It was suggested that reduced availability of VLCPUFA in the diet led to efficient retention and redistribution of DHA [48]. More recent studies with echium oil in salmon [49] and barramundi [50] failed to show any major benefits of SDA-enriched oils in fish feeding trials. It is worthwhile noting, however, that the SDA content of echium (14%) is relatively low compared to what may be obtainable from a transgenic plant source, and the presence of GLA as competing ω-6 byproduct may also affect the conversion efficiencies observed in feeding trials. Thus, a plant source high in SDA and low in GLA and other ω-6 PUFA may be more effective at shifting fish metabolism toward conversion of ω-3 PUFA.

Given that the flesh fatty acid composition typically reflects the lipid composition of the diet, it becomes clear that direct substitution of fish oil with EPA and DHA-enriched oils may be the most effective means of meeting the ω-3 VLCPUFA requirements of farmed fish. As described earlier, efforts to produce a sustainable land-based source of EPA and DHA are advancing rapidly; however, the intensifying pressure on the global fish oil supply requires a multifaceted approach to address the sustainability of the aquaculture sector. To that end, the use of different feeding phases in fish production has been explored as a way to make the most efficient use of available dietary lipids (reviewed by [35]). These trials use different starter, grow-out, and finishing feeds to address the different dietary needs at various stages of development. With respect to fish development, the need for dietary VLCPUFA (i.e., fish oil) is greatest during the starter phase, in which rapid growth, high survival, and normal development (particularly of neural tissue) are the most important considerations. Finishing feeds, though not yet common in commercial settings, may also call for a greater VLCPUFA content in order to ensure that the desired postharvest fatty acid composition is obtained.

There seems to be the greatest potential for fish oil substitution during the grow-out phase, during which the flesh fatty acid composition is not as critical and growth rate and feed conversion are the main considerations. In this case, a VLCPUFA-rich finishing diet would facilitate "washing-out" of the less desirable flesh FA prior to harvest. Such feeding strategies would allow fish producers to reduce their overall reliance on fish oil through more efficient feeding strategies, thereby reducing feed costs and enhancing the sustainability of the operation.

## 7.4    OILSEED PROCESSING TECHNOLOGIES FOR AQUAFEED

There has been considerable progress toward developing sustainable alternatives to fish meal and fish oil for aquafeed applications [51]. Today, the pork, beef, and poultry industries are the major users of oilseed meals for feed applications, but advancements in processing technology may expand market opportunities into the aquaculture sector. While the development of specialty oils containing high levels of physiologically important PUFA remains a major focus, advancements in oilseed processing technologies are also crucial for the production of high-quality protein concentrates for fish nutrition. For example, plant seeds contain a substantial amount of fiber, which is considered undesirable for fish feed. Novel processes based on fractionation of the oilseed cake or meal may facilitate the removal of antinutritional components left behind by traditional processing [52] and yield a product more amenable for use in fish feed.

The interest in processing technologies has been driven in part by the growth of so-called biorefineries, which aim to maximize the value of all components of a particular feedstock. The rapid expansion of oilseed-based biofuels has sparked increased interest in finding alternative markets for the leftover seed meal, which would help offset the relatively high feedstock costs involved in biodiesel production. Technologies that recover high-value components from the meal or otherwise increase the value of the meal for feed applications would substantially broaden the operating margins for biorefining facilities while helping to meet the needs of specialty markets such as the aquaculture sector.

The typical industrial process for production of edible oils involves pressing the seeds, followed by a solvent extraction, usually with hexane. This method results in a good recovery of high-quality, low cost oil, with residual seed meal being used by the animal feed industry. However, environmental considerations, as well as the desire to increase the value of the meal, have led researchers to explore processing technologies that rely less on organic solvents and take advantage of aqueous or enzyme-assisted technologies. Enzymes can be used to facilitate oil extraction by hydrolyzing the cell wall polysaccharides and oilbody membranes, and have the advantage of yielding nondenatured protein, which has higher value as a protein concentrate (containing fiber) or protein isolate (fiber removed). Advancements in industrial enzyme production are making enzym e-assisted processes increasingly cost-efficient, and several modifications of aqueous processes have been described in papers and patents [52, 53]. Several companies have already patented technologies for production of canola and flax protein isolates, and industrial pilot plants using aqueous technologies are already in operation.

Several processes for the manufacture of protein concentrates or isolates for aquaculture have also been developed. These technologies yield nondenatured proteins with better digestibility for fish nutrition. Membrane separation techniques are used to separate oilseed components, remove antinutritional factors, and fractionate protein for fish feed production.

VLCPUFA-enriched oilseeds pose additional challenges for processing since the high PUFA content of the oils reduces their thermal and oxidative stability. Aqueous milling processes, followed by separation of the lipid fraction, are well-suited for the preservation of the high-value VLCPUFA, since the harsh refining steps (bleaching, degumming and deodorization) are not needed for fish feed production. The presence of phospholipids is actually considered beneficial for fish nutrition.

Supercritical fluid extraction is another possible strategy for processing of oilseeds for aquafeed applications, but so far the expansion of this technology has been inhibited by the cost of the process and lack of equipment for continuous processing using this technology [54].

## 7.5   CONCLUSION

Aquaculture production has grown immensely since the mid-1960s, and has now become the world's fastest growing primary food production sector, providing almost half of all the fish consumed by humans today [55]. At the same time, production from capture fisheries has declined due to chronic overfishing – largely to support the expansion of the global aquaculture industry. In 2006, global aquaculture used 68% of fish meal and 88% of fish oil produced worldwide [56], with a few carnivorous species (e.g., trout and salmon) consuming 19.5% and 51% of meal and oil, respectively, despite the fact that they represent only 3% of global aquaculture production [56].

Clearly, neither capture fisheries nor aquaculture can sustainably meet the needs of a human population expected to exceed 9 billion by 2050, unless the aquaculture industry can overcome its reliance on capture fisheries as a source of feed ingredients. Future growth of the industry depends on the development of alternative feed ingredients, a large fraction of which are likely to come from terrestrial plants, particularly from modified oilseeds containing EPA and DHA. Advancements in oilseed biotechnology and processing have set the stage for such designer oilseeds to make a major impact on the aquafeed market in the years to come.

## KEYWORDS

- acyl-editing
- aquafeed
- Arabidopsis
- Brassica carinata
- Brassica juncea
- Carthamus tinctorius
- Gadus morhua
- Onchorhynchous tshawytscha
- Salvelinus alpines

## REFERENCES

1. Subasinghe, R.; Soto, D.; Jia, J: global aquaculture and its role in sustainable development. *Rev Aquaculture* **2009**, 1: 2–9.
2. de Silva, S.; Francis, D.; Tacon, A. fish oils in aquaculture. In *Fish Oil Replacement and Alternative Lipid Sources in Aquaculture Feeds*. CRC Press; 2010: 1–20.
3. Simopoulos, A. P. the importance of the omega-6/omega-3 fatty acid ratio in cardiovascular disease and other chronic dizeases. *Exp Biol Med (Maywood)* **2008**, 233: 674–688.
4. Venegas-caleron, M.; Sayanova, O, Napier, J. A: an alternative to fish oils: metabolic engineering of oil-seed crops to produce omega-3 long chain polyunsaturated fatty acids. *Prog Lipid Res* **2010**, 49: 108–119.
5. Nykiforuk, C. L.; Shewmaker, C, Harry I, Yurchenko, O. P.; Zhang, M, Reed, C *et al.*: high level accumulation of gamma linolenic acid (C18: 3Delta6.9, 12 cis) in transgenic safflower (*Carthamus tinctorius*) seeds. *Transgenic Res* **2011.**
6. Cheng, B.; Wu, G.; Vrinten, P.; Falk, K.; Bauer, J.; Qiu, X.;: towards the production of high levels of eicosapentaenoic acid in transgenic plants: the effects of different host species, genes and promoters. *Transgenic Res* **2010**, 19: 221–229.
7. Gatlin, D. M.; Barrows, F. T.; Brown P, Dabrowski, K.; Gaylord, T. G.; Hardy RW *et al.*: expanding the utilization of sustainable plant products in aquafeeds: a review. *Aquaculture Res* **2007**, 38: 551–579.
8. Plourde, M.; Cunnane, S. C. extremely limited synthesis of long chain polyunsaturates in adults: implications for their dietary essentiality and use as supplements. *Appl Physiol Nutr Metab* **2007**, 32: 619–634.
9. Griffiths, G, Morse, N: clinical Applications of C-18 and C-20 Chain Length Polyunsaturated Fatty Acids and Their Biotechnological Production in Plants. *J Am Oil Chem Soc* **2006**, 73: 171–185.

10. Uttaro, AD: biosynthesis of polyunsaturated fatty acids in lower eukaryotes. *IUBMB Life* **2006,** 58: 563–571.
11. Metz, J. G.; Roessler, P, Facciotti, D, Levering, C, Dittrich, F, Lassner, M *et al.*: production of polyunsaturated fatty acids by polyketide synthases in both prokaryotes and eukaryotes. *Science* **2001,** 293: 290–293.
12. Griffiths, G, Stobart, A. K.; Stymne, S: delta 6- and delta 12-desaturase activities and phosphatidic acid formation in microsomal preparations from the developing cotyledons of common borage (*Borago officinalis*). *Biochem J* **1988,** 252: 641–647.
13. Tocher, D. R. fatty acid requirements in ontogeny of marine and freshwater fish. *Aquaculture Res* **2010,** 41: 717–732.
14. Hong, H.; Datla, N.;Reed, D. W.; Covello, P. S.; MacKenzie, S. L.; Qiu X. high-level production of gamma-linolenic acid in *Brassica juncea* using a delta6 desaturase from *Pythium irregulare*. *Plant Physiol* **2002,** 129: 354–362.
15. Liu, J. W.; DeMichele, S.; Bergana, M.; Bobik, E.; Hastilow, C.; Chuang, L. T. *et al.*: characterization of oil exhibiting high gamma-linolenic acid from a genetically transformed canola strain. *J Am Oil Chem Soc* **2001,** 78: 489–493.
16. Eckert, H, La VB, Schweiger, B. J.; Kinney, A. J.; Cahoon, E. B.; Clemente T: coexpression of the borage Delta 6 desaturase and the Arabidopsis Delta 15 desaturase results in high accumulation of stearidonic acid in the seeds of transgenic soybean. *Planta* **2006,** 224: 1050–1057.
17. Ruiz-lopez, N.; Haslam, R. P.; Venegas-caleron M.; Larson, T. R.; Graham, I. A.; Napier J. A. *et al.*: the synthesis and accumulation of stearidonic acid in transgenic plants: a novel source of 'heart-healthy' omega-3 fatty acids. *Plant Biotechnol J* **2009,** 7: 704–716.
18. Bell, J. G.; Strachan F.; Good, J. E.; Tocher. D.R. effect of dietary echium oil on growth, fatty acid composition and metabolism, gill prostaglandin production and macrophage activity in Atlantic cod (*Gadus morhua* L.). *Aquaculture Res* **2006,** 37: 606–617.
19. Abbadi, A.; Domergue F; Bauer J.; Napier, J. A.; Welti R, Zahringer U *et al.*: biosynthesis of very-long-chain polyunsaturated fatty acids in transgenic oilseeds: constraints on their accumulation. *Plant Cell* **2004,** 16: 2734–2748,
20. Domergue, F, Abbadi A, Ott C, Zank, T. K.; Zahringer U, Heinz E: acyl carriers used as substrates by the desaturases and elongasses involved in very long-chain polyunsaturated fatty acids biosynthesis reconstituted in yeast. *J Biol Chem* **2003,** 278: 35115–35126.
21. Qi, B.; Fraser, T.; Mugford, S.; Dobson, G.; Sayanova, O. Butler J. *et al.*: production of very long chain polyunsaturated omega-3 and omega-6 fatty acids in plants. *Nat Biotechnol* **2004,** 22: 739–745.
22. Wu, G, Truksa M, Datla N, Vrinten, P, Bauer J, Zank *t et al.*: stepwize engineering to produce high yields of very long-chain polyunsaturated fatty acids in plants. *Nat Biotechnol* **2005,** 23: 1013–1017.
23. Kajikawa, M.; Matsui, K.; Ochiai M.; Tanaka, Y.; Kita Y.; Ishimoto, M. *et al.*: production of arachidonic and eicosapentaenoic acids in plants using bryophyte fatty acid Delta6-desaturase, Delta6-elongase, and Delta5-desaturase genes. *Biosci Biotechnol Biochem* **2008,** 72: 435–444.

24. Hoffmann, M.; Wagner, M.; Abbadi, A.; Fulda, M.; Feussner I: metabolic engineering of omega-3-very long chain polyunsaturated fatty acid production by an exclusively acyl-coA-dependent pathway. *J Biol Chem* **2008**, 283: 22352–22362.
25. Dahlqvist, A.;Stahl, U; Lenman, M.; Banas, A.; Lee, M.; Sandager, L. *et al.*: phospholipid: diacylglycerol acyltrans-ferase: an enzyme that catalyzes the acyl-coA-independent formation of triacylglycerol in yeast and plants. *Proc Natl Acad Sci U S A* **2000**, 97: 6487–6492.
26. Lu, C.F.; Xin , Z.G.; Ren, Z. H.; Miquel, M.; Browse J: an enzyme regulating triacylglycerol composition is encoded by the ROD1 gene of Arabidopsis. *Proc Natl Acad Sci USA* **2009**, 106: 18837–18842.
27. Bates, P. D.; Durrett, T. P.; Ohlrogge, J. B.; Pollard, M. analysis of acyl fluxes through multiple pathways of triacylglycerol synthesis in developing soybean embryos. *Plant Physiol* **2009**, 150: 55–72.
28. Bates, P. D.; Browse, J. The pathway of triacylglycerol synthesis through phosphatidylcholine in *Arabidopsis* produces a bottleneck for the accumulation of unusual fatty acids in transgenic seeds. *Plant J* **2011**.
29. Wood, C. C.; Petrie, J. R.; Shrestha P, Mansour, M. P.; Nichols, P. D.; Green A.G. *et al.*: a leaf-based assay using interchangeable design principles to rapidly assemble multistep recombinant pathways. *Plant Biotechnol J* **2009**, 7: 914–924.
30. Petrie, J. R.; Shrestha, P.; Liu, Q.; Mansour, M. P.; Wood, C. C.; Zhou X.R. *et al.*: rapid expression of trans-genes driven by seed-specific constructs in leaf tissue: dHA production. *Plant Methods* **2010**, 6: 8.
31. Barker B.; Brain, eye-, and heart-healthy canola oil in the works. Top Crop Manager (http:,www.topcropmanager.com/content/view/4128/38/). **2008**, 5–15–2011. Ref Type: electronic Citation
32. Nuseed. Australian scientific collaboration set to break world's reliance on fish for long chain omega-3. Nuseed media release. http:,www.nuseed.com/Assets/448/1/110412_Omega3inCanola_PioneeringAustralianresearchallianceannouncement_Embargo-tuesday12April2011,pdf. **2011**, 5–15–2011, Ref Type: electronic Citation
33. Watkins, C. Oilseeds of the future: part 1. INFORM (http:,www.aocs.org/Membership/FreeCover.cfm?itemnumber=1086). **2009**, 5–15–2011, Ref Type: electronic Citation
34. Monsanto. World's first SDA omega-3 soybean oil achieves major mileston that advances the development of foods with enhanced nutritional benefits. http:,monsanto.mediaroom.com/index.php?s=43anditem=761andprintable. **2009**, 5–15–2011, Ref Type: electronic Citation
35. Glencross, B.; Turchini, G.; fish oil replacement in starter, grow-out, and finishing feeds for farmed aquatic animals. In *Fish Oil Replacement and Alternative Lipid Sources in Aquaculture Feeds*. CRC Press; 2010: 373–404.
36. Turchini, G. M.; Torstensen, B. E.; Ng WK: fish oil replacement in finfish nutrition. *Rev Aquaculture* **2009**, 1: 10–57.
37. Das U.N.; essential fatty acids – a review. *Curr Pharm Biotechnol* **2006**, 7: 467–482.
38. Mugrditchian, D. S.; Hardy, R. W.; Iwaoka. W.T.; linseed oil and animal fat as alternative lipid sources in dry diets for Chinook Salmon (*Oncorhynchus tshawytscha*). *Aquaculture* **1981**, 25: 161–172.

39. Stubhaug I.; Lie O.; Torstensen B.E.; fatty acid productive value and beta-oxidation capacity in Atlantic salmon (*Salmo salar* L.) fed on different lipid sources along the whole growth period. *Aquaculture Nutr* **2007**, 13: 145–155.

40. Menoyo D.; Lopez-bote, C. J..; Diez A.; Obach A.; Bautista J.M.; impact of n-3 fatty acid chain length and n-3/n-6 ratio in Atlantic salmon (*Salmo salar*) diets. *Aquaculture* **2007**, 267: 248–259.

41. Greene D.H.S.; Selivonchick D. P.; effects of dietary vegetable, animal and marine lipids on muscle lipid and hematology of rainbow-trout (*Oncorhynchus mykiss*). *Aquaculture* **1990**, 89: 165–182.

42. Yamazaki K.; Fujikawa M.; Hamazaki T.; Yano S.; Shono T.; comparison of the conversion rates of alpha-linolenic acid (18–3(N-3)) and stearidonic Acid (18–4(N-3)) to longer polyunsaturated fatty acids in rats. *Biochimica et Biophysica Acta* **1992**, *1123*, 18–26.

43. James, M. J.; Ursin, V. M.; Cleland L.G. metabolism of stearidonic acid in human subjects: comparison with the metabolism of other n-3 fatty acids. *Am J Clin Nutr* **2003**, *77*, 1140–1145.

44. Harris, W. S.; Lemke, S. L.; Hansen, S. N.; Goldstein, D. A.; DiRienzo, M. A.; Su H *et al.*: stearidonic acid-enriched soybean oil increased the omega-3 index, an emerging cardiovascular risk marker. *Lipids* **2008**, *43*, 805–811.

45. Brenna, J. T.; Salem, N.; Sinclair, A. J.; Cunnane SC: alpha-linolenic acid supplementation and conversion to n-3 long-chain polyunsaturated fatty acids in humans. *Prostag Leukotr Ess* **2009**, *80*, 85–91.

46. Tocher, D. R.; Dick, J. R.; MacGlaughlin P, Bell J G. effect of diets enriched in Delta 6 desaturated fatty acids (18: 3n-6 and 18: 4n-3), on growth, fatty acid composition and highly unsaturated fatty acid synthesis in two populations of Arctic charr (*Salvelinus alpinus* L.). *Comp Biochem Phys B* **2006**, *144*, 245–253.

47. Miller, M. R.; Nichols, P. D.; Carter C. G. replacement of dietary fish oil for Atlantic salmon parr (*Salmo salar* L.) with a stearidonic acid containing oil has no effect on omega-3 long-chain polyunsaturated fatty acid concentrations. *Comp Biochem Phys B* **2007**, *146*, 197–206.

48. Tocherl D.; Francis, D.; Coupland, K. N-3 Polyunsaturated fatty acid-rich vegetable oils and blends. In *Fish Oil Replacement and Alternative Lipid Sources in Aquaculture Feeds*. CRC Press; **2010**, 209–244.

49. Miller, M. R.; Bridle, A. R.; Nichols, P. D.; Carter C.G. Increased elongase and desaturase gene expression with stearidonic acid enriched diet does not enhance long-chain (n-3) content of seawater Atlantic Salmon (*Salmo salar* L.). *J Nutr* **2008**, *138*, 2179–2185.

50. Alhazzaa, R.; Bridle, A. R.; Nichols, P. D.; Carter C. G. Replacing dietary fish oil with Echium oil enriched barramundi with C-18 PUFA rather than long-chain PUFA. *Aquaculture* **2011**, *312*, 162–171.

51. de Silva, S.; Soto, D. climate change and aquaculture: potential impacts, adaptation and mitigation. *FAO Fisheries Technical Paper* **2009**, *530*, 137–215.

52. Muellerl, K.; Eisner, P.; Yoshie-stark, Y.; Nakada, R.; Kirchhoff, E. functional properties and chemical composition of fractionated brown and yellow linseed meal (*Linum usitatissimum* L.). *J. Food Engineering* **2010**, *98*, 453–460.

53. Nennerfelt, L.; Evans, D.; Ten Haaf, W. Process for extracting flax protein concentrate from flax meal. Nutrex Wellness, Inc. [US 6998466 B2]. **2006,** Ref Type: patent.
54. Brunner, G. Supercritical fluids: technology and application to food processing. *J. Food Engineering* **2005,** *67,* 21–33.
55. Food and Agriculture Organization. The status of world fisheries and aquaculture. **2009,** Rome, Food and Agriculture Organization. Ref Type: report.
56. Tacon, A.G.J.; Metian, M. Global overview on the use of fish meal and fish oil in industrially compounded aquafeeds: trends and future prospects. *Aquaculture* **2008,** *285,* 146–158.

# CHAPTER 8

# MICROBIAL BIOSENSORS

ANAMIKA SINGH, and RAJEEV SINGH

## CONTENTS

## 8.1   INTRODUCTION

A biosensor is an automatic device, which can help us by transforming biological (enzymatic) response to electrical in the form of fluorescent light. The emphasis of this topic concerns enzymes as the biologically responsive material, but it should be recognized that other biological systems may be used by biosensors, for example, whole cell metabolism, Ligand binding and the antibody-antigen reaction. Biosensors represent a rapidly expanding field, at the present time, with an estimated 60% annual growth rate; the major impetus coming from the health-care industry (e.g., 6% of the western world are diabetic and would benefit from the availability of a rapid, accurate and simple biosensor for glucose) but with some pressure from other areas, such as food quality appraisal environmental monitoring and defense purposes.

Before designing biosensors we have to make sure about the following important feature must be considered:

1.   The biocatalyst should be specific for particular purpose.
2.   The reaction should be independent of parameters like pH and stirring etc.
3.   The response should not be affected with concentration or dilution and it should be free from electrical noise.
4.   Biosensor, which is used for clinical situations, the probe must be biocompatible and having no toxic or antigenic effects. The biosensor should not be fouling or proteolysis.
5.   The complete biosensor should be cheap, small, portable and capable of being used as operators.
6.   There should be a market for the biosensor.

In general biosensors must have five components: biocatalyst, Transducer, Amplifier, Processor and Displaying devices. Biocatalysts are made by using immobilization techniques. The most common immobilization techniques are physical adsorption, cross-liking, entrapment, covalent bond, or some combination of all of these techniques the stability of the immobilization techniques determines the sensitivity and reliability of the biosensor signal. Transducers, required pieces of analytical tools that convert the biological and chemical changes into the useful electronic data, make use of these immobilization techniques to provide different types of transducers, for example, optical, piezoelectric, etc. The analyst selective interface is a biological active substance such as enzyme, antibody, DNA (deoxyribonucleic acid) and microorganism. Such substances are cable of recognizing their specific analyzes and regulating the overall performance (response time, reliability, specificity, selectivity and sensitivity) of the biosensors. Biosensors are analytical tools that consist of a substrate and a selective interface in closed proximity or integrate with transducer; therefore, the substrates and transducers are important components of the analytical tools, which contain an immobilized biologically active compound that can interact with specific species (substrates) produces a products in the form of a biological or chemical substance, heat, light, or sound: then a transducer such as a electrode, semiconductor, thermistor, photocounter or sound detector changes the product of the reaction into usable data. The data will be further amplified in the presence of reference data and then data will be processed and displayed in the user-friendly manner (Fig. 1). In this chapter we are giving more emphasis on microbes as biosensors. Microbes used in biosensors can also be called as Bioreporters.

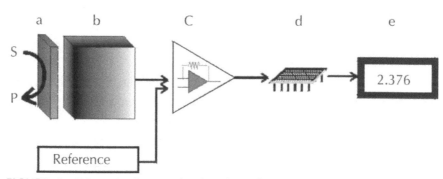

**FIGURE 1**    Schematic diagram showing the main components of a biosensor. The biocatalyst (a) converts the substrate to product. This reaction is determined by the transducer, which (b) converts it to an electrical signal. The output from the transducer is amplified (c), processed (d) and displayed (e). (http://vanibala.150 m.com).

Bioreporters refer to intact, live microbial cells, which are genetically engineered to produce a detectable signal in response to particular agent in their environment (Fig. 2). Bioreporters are having two essential genetic elements, a promoter gene and a reporter gene. The promoter gene is turned on (transcribed) when the target agent is present in the cells environment. The promoter gene in a normal bacterial cell is linked to other genes that are then likewise transcribed and then translated into proteins that help the cell in either combating or adapting to the agent to which it has been exposed. In the case of a bioreporter, these genes, or portions thereof, have been removed and replaced with a reporter gene. Consequently, turning on the promoter gene now causes the reporter gene to be turned on. Activation of the reporter gene leads to production of reporter proteins that ultimately generate some type of a detectable signal.

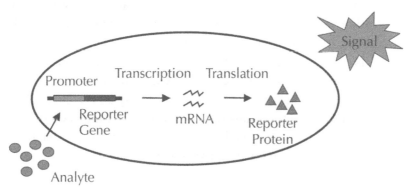

**FIGURE 2**   Steps in Biorepoter. Upon exposure to specific analyse, the promoter/ reporter gene complex is transcribed into messenger RNA (mRNA) and then translated into reporter protein that is ultimately responsible for generating a signal. (Curtsey: http://en.wikipedia. org/wiki/Bioreporter).

Therefore, the presence of a signal indicates that the bioreporter has sensed a particular target agent in its environment.

Originally developed for fundamental analysis of factors affecting gene expression, bioreporters were early on applied for the detection of environmental contaminants1 and have since evolved into fields as diverse as medical diagnostics, precision agriculture, food-safety assurance, process monitoring and control, and bio-microelectronic computing. Their versatility stems from the fact that there exist a large number of reporter gene systems that are capable of generating a variety of signals. Additionally, reporter genes can be genetically inserted into bacterial, yeast, plant, and mammalian cells, thereby providing considerable functionality over a wide range of host vectors.

## 8.2   REPORTER GENE SYSTEMS

Several types of reporter genes are available for use in the construction of bioreporter organisms, and the signals they generate can usually be categorized as either colorimetric, fluorescent, luminescent, chemiluminescent or electrochemical. Although each functions differently, their end product always remains the same- a measurable signal that is proportional to the concentration of the unique chemical or physical agent to which they have been exposed. In some instances, the signal only occurs when a secondary substrate is added to the bioassay (*luxAB*, Luc, and aequorin). For other bioreporters, the signal must be activated by an external light source (GFP and UMT), and for a select few bioreporters, the signal is completely self-induced, with no exogenous substrate or external activation being required (*luxCDABE*). The following sections outline in brief some of the reporter gene systems available and their existing applications.

### 8.2.1   BACTERIAL LUCIFERASE (LUX)

Luciferase is a generic name for an enzyme that catalyzes a light-emitting reaction. Luciferases enzyme founds in bacteria, algae, fungi, jellyfish, insects, shrimp, and squid. In bacteria, light emitting gene have been isolated and used for light emitting reactions for different organisms. Bioreporters that emit a blue-green light with a maximum intensity at 490 nm (Figure 3) Three variants of *lux* are available, one that functions at <30°C, another at <37°C, and a third at <45°C. The *lux* genetic system consists of five genes, *luxA*, *luxB*, *luxC*, *luxD*, and *luxE*. Depending on the combination of these genes used, several different types of bioluminescent bioreporters can be constructed.

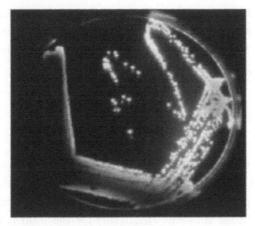

**FIGURE 3**   Genetically engineered bacteria with lux gene glows in culture.

### 8.2.2   LUXAB BIOREPORTERS

luxAB bioreporters are actually *luxA* and *luxB* genes, which together generat the light signal. However, to fully complete the light-emitting reaction, a substrate must be supplied to the cell. *luxAB* bioreporters have been constructed within bacterial, yeast, insect, nematode, plant, and mammalian cell systems. Few bioreporters having luxAB are bacterial biofilm, bacterial biomass, cell viable counts which perform different functions.

### 8.2.3   LUXCDABE BIOREPORTERS

Bioreporters can contain all five genes of the *lux* cassette, thereby allowing for a completely independent light generating system that requires no extraneous additions of substrate nor any excitation by an external light source. So in this bioassay, the bioreporter is simply exposed to a target analyte and a quantitative increase in bioluminescence results, often within less than one hour. Due to their rapidity and ease of use, along with the ability to perform the bioassay repetitively in realtime and on-line, makes *luxCDABE* bioreporters extremely attractive. Consequently, they have been incorporated into a diverse array of detection methodologies ranging from the sensing of environmental contaminants to the real-time monitoring of pathogen infections in living mice. e.g., there are few Lux CDABE based bioreporter which detect 2,3 Dichlorophenol,2,4,6, Trichlorophenol, 2,4, D,3- Xylene etc.

## 8.2.4  FIREFLY LUCIFERASE

Firefly luciferase (Luc) catalyzes a reaction that produces visible light in the 550–575 nm range. A click-beetle luciferase is also available that produces light at a peak closer to 595 nm. Both luciferases require the addition of an exogenous substrate (luciferin) for the light reaction to occur. Numerous *luc*-based bioreporters have been constructed for the detection of a wide array of inorganic and organic compounds of environmental concern.

## 8.2.5  AEQUORIN

Aequorin is a photoprotein isolated from the bioluminescent jellyfish *Aequorea victoria*. Upon addition of calcium ions (Ca2+) and coelenterazine, a reaction occurs whose end result is the generation of blue light in the 460–470 nm range.

## 8.2.6  GREEN FLUORESCENT PROTEIN

Green fluorescent protein (GFP) is also a photoprotein isolated and cloned from the jellyfish *Aequorea victoria*. Variants have also been isolated from the sea pansy *Renilla reniformis*. GFP, like aequorin, produces a blue fluorescent signal, but without the required addition of an exogenous substrate. All that is required is an ultraviolet light source to activate the fluorescent properties of the photoprotein. This ability to autofluoresce makes GFP highly desirable in biosensing assays since it can be used on-line and in real-time to monitor intact, living cells. Additionally, the ability to alter

GFP to produce light emissions besides blue (i.e., cyan, red, and yellow) allows it to be used as a multianalyte detector.

Consequently, GFP has been used extensively in bioreporter constructs within bacterial, yeast, nematode, plant, and mammalian hosts (Fig. 4). Some examples of GFP applications in mammalian cell systems, where its use has revolutionized much of what we understand about the dynamics of cytoplasmic, cytoskeletal, and organellar proteins and their intracellular interactions. Some other applications of GFP are monitoring tumor cells in gene therapy protocols, identification of HIV in infected cells and tissue, monitoring of protein – protein interactions in living cells, etc.

## 8.2.7 UROPORPHYRINOGEN (UROGEN) III METHYLTRANSFERASE (UMT)

uMT catalyzes a reaction that yields two fluorescent products which produce a red-orange fluorescence in the 590–770 nm range when illuminated with ultraviolet light. So as with GFP, no addition of exogenous substrates is required.

## 8.3 DETECTING THE OPTICAL SIGNATURE

Using light as the terminal indicator is advantageous in that it is an easily measured signal. Optical transducers such as photo multiplier tubes, photodiodes, micro channel plates, or charge-coupled devices are readily available and can be easily integrated into high throughput readers. As these usually consist of large, tabletop devices, demand for smaller, portable light readers for remote monitoring has resulted in the development of battery-operated, hand-held photo multiplier units. Recently, The Center for environmental Biotechnology and Oak Ridge National Laboratory (ONRL) have taken steps towards genuine miniaturization of optical transducers and have successfully developed integrated circuits capable of detecting bioluminescence directly from bioreporter organisms.

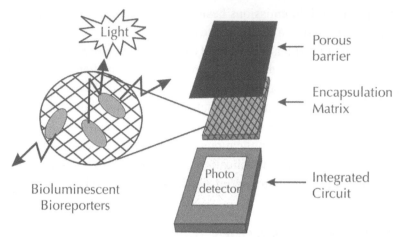

**FIGURE 4**   Assembly of a bioluminescent bioreporter integrated circuit (BBIC) sensor.

These bioluminescent bioreporter integrated circuits (BBICs) consist of two main components; photo-detectors for capturing the on-chip bioluminescent bioreporter signals and signal processors for managing and storing information derived from bioluminescence (Fig. 4). Remote frequency (RF) transmitters can also be incorporated into the overall integrated circuit design for wireless data relay. Since the bioreporter and biosensing elements are completely self-contained within the BBIC, operational capabilities are realized by simply exposing the BBIC to the desired test sample.

In addition to incorporation in a BBIC format, the whole-cell bioreporter matrix can also be immobilized on something as simple as an indicator test strip. In this fashion, a home water quality indicator, for example, could be developed to operate in much the same way as a home pregnancy test kit.

Genetically modified microbes are used extensively in various fields like Defense, Medical, Environment, Food and Agriculture etc.

## 8.4   BIOSENSORS IN THE FIELD OF DEFENSE

### 8.4.1   LAND MINES

Every year in world, due to land mines accidents more than 25,000 people have been killed. Land mines are extremely difficult to find once they are buried in the ground. Plastic mines are almost impossible to locate because they elude metal detectors. Fortunately, most land mines leak slightly and leave traces of explosive chemicals such as TNT shortly after they are installed. ORNL has developed a clever way of using bacteria to detect this faint explosive signature.

Bob Burlage, a microbiologist in ORNL's (Environmental Sciences Division) has genetically engineered microorganisms to emit light in the presence of TNT. As they recognize and consume TNT, the engineered bacteria produce a fluorescent protein that appears as a green light when they are illuminated by ultraviolet (UV) light. Landmines leaks small amount of TNT over time (Fig. 5)

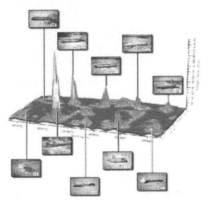

**FIGURE 5** Land mines mapping by using after Microbial biosensors.

When the bacterial strain of *Pseudomonas putida* encounter the TNT, they will scavenge the compound as a food source, activating the genes that produce proteins needed to digest the TNT, because green fluorescent protein gene obtained from jellyfish to these activated genes and included a regulatory gene that recognizes TNT. As a result, the attached gene will also be turned on. It will produce the green fluorescent protein, which emits extremely bright fluorescence when exposed to UV light (Fig. 6). It is one of the most rapid and advanced methods for landmines detection. The field has been sprayed with bacteria during the day using a agriculture sprayer. At night rolling towers and helicopters by looking for glowing microbes on soil illuminated UV light, ONGL will get the green light to develop a bacteria remote sensing method for land mines detection. This technique offers several advantages. It is inexpensive, it poses no hazard to operators and it is virtually the only mine detector technology which wide areas of fields.

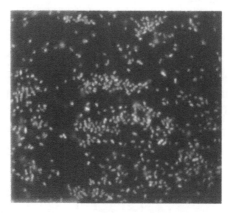

**FIGURE 6**    Microbial biosensors with GFP illuminated with UV light.

## 8.4.2    BIOSNIFFING

Biosniffing another development in biosensors, its nothing but recognizing the airborne pathogens by using sensors. The major defense concern in USA after September 9/11 is given to control bioterrorism by detecting airborne pathogens like bubonic plague, anthrax, botulism, small pox, tularemia etc. by using wireless Biosensors [1].

## 8.4.3    SPACE

At the University of Tennessee's Center for Environmental Biotechnology, researchers are developing wireless biosensors that can be used to monitor microbial contamination and radiation exposure in future manned space missions. The researchers are using "bioluminescent biosensor organisms" built directly onto wireless integrated circuits. They call them Bioluminescent Bioreporter Integrated Circuits (BBIC). Bioluminescent biosensors are bacteria that glow when they come into contact with a specific toxic compound. The more poison the bacteria receive, the more brightly they shine. The BBIC uses this light to determine the level of toxicity present in the spacecraft, sending a wireless message to a central computer. Researchers say, "The ultimate goal is to create an array or network of small, unobtrusive, low cost, low power BBICs for intelligent distributed monitoring of the space craft environment, as well as for planetary-based surface habitats." It may sound like science fiction, but this is one technology that's coming down to earth, and not a moment too soon.

## 8.5   BIOSENSORS IN THE FIELD OF MEDICAL

### 8.5.1   CHONDROCYTES

One of the main open questions in chondrocyte transplantation is the fate of the implanted cells in vivo. Intended to establish prerequisites for such studies in animal models and to show the feasibility of this approach in rabbits. Isolated articular chondrocytes were retro virally marked using green fluorescence protein (GFP) as a cell-specific marker in order to allow an in vivo follow-up of these cells [2].

### 8.5.2   CANCER

Recently, there has been tremendous interest in developing techniques such as MRI, microCT, microPET, and SPECT to image function and processes in small animals. These technologies offer deep tissue penetration and high spatial resolution, but compared with noninvasive small animal optical imaging, these techniques are very costly and time consuming to implement. Optical imaging is cost-effective, rapid, easy to use, and can be readily applied to studying disease processes and biology in vivo. In vivo optical imaging is the result of a coalescence of technologies from chemistry, physics, and biology. The development of highly sensitive light detection systems has allowed biologists to use imaging in studying physiological processes. Over the last few decades, biochemists have also worked to isolate and further develop optical reporters such as GFP, luciferase, and cyanine dyes. This article reviews the common types of fluorescent and bioluminescent optical imaging, the typical system platforms and configurations, and the applications in the investigation of cancer biology [3].

In cancer diagnosis the Insertion of the *luc* genes into a human cervical carcinoma cell line (HeLa) illustrated that tumor-cell clearance could be visualized within a living mouse by simply scanning with a charge-coupled device camera, allowing for chemotherapy treatment to rapidly be monitored on-line and in real-time4. In another example, the *luc* genes were inserted into human breast cancer cell lines to develop a bioassay for the detection and measurement of substances with potential estrogenic and antiestrogenic activity.

Optical imaging is a modality that is cost-effective, rapid, easy to use, and can be readily applied to studying disease processes and biology in vivo. For this study, we used a green fluorescent protein (GFP)– and luciferase-expressing mouse tumor model to compare and contrast the quantitative and qualitative capabilities of a fluorescent reporter gene (GFP) and a bioluminescent reporter gene (luciferase) [4]. Biosensors in medical field grow extensively from previous few years ornithine decarboxylase and *S. adenosylmenthionine* decarboxylase (AdometDC) are two enzymes of polyamine biosynthesis in cells induces metagenesis in cells which leads cancer development, ODC inhibitor DFMO induces polyamine depletion result cytostatic growth inhibitor [5].

### 8.5.3   HIV (AIDS)

Highly active retroviral therapy (HAART) which consists of inhibitor of viral enzyme (reverse transcriptase (RT) and proteases) which is also a part of biosensor, i.e., induction of inhibitor viral enzyme into host and its use as biosensor against HIV [6]. In field of medicine there are nucleic acid biosensor which helps in medical diagnosis [7].

## 8.5.4  HUMAN B CELL LINE

Aequorin has been incorporated into human B cell lines for the detection of pathogenic bacteria and viruses in what is referred to as the CANARY assay (Cellular Analysis and Notification of Antigen Risks and Yields). The B cells are genetically engineered to produce aequorin. Upon exposure to antigens of different pathogens, the recombinant B cells emit light as a result of activation of an intracellular signaling cascade that releases calcium ions inside the cell [8].

## 8.6  BIOSENSORS IN THE FIELD OF ENVIRONMENT

Heavy metals are an important class of environmental hazards, and as the heavy metals in industry continues to increase, larger segments of biota including human beings, will be exposed to increasing levels of the toxicants, detection of individual metals in the environment may event be possible using biosensor consisting of genetically engineered microorganism [9]. Genetically engineered *Pseudomonas species* (Shk1) with bioluminescence capacity isolated from activated sludge, are able to report or detect heavy metal toxicity [10].

Mainly *Pseudomonas fluorescence* and bakers yeast (*Saccharomyces cerevisiae*) were used for heavy metal detection. Lac-z based reporter gene transposon Tn5B20 was performed random gene fusions in the genome of the common soil bacteria. *Pseudomonas fluorescence* strains ATCC13525 were used to create a bank of 5,000 *P. fluorescence* mutants. This mutant bank in heavy metal detection was screened for differential gene expression in the presence of toxic metal Cadmium. One of mutant c-8 shows increase gene expression in the presence of ethanol. Mutant c11 is hypersensitive, to cadmium and zinc ions, it has a sensor/regulator protein pair which is also found in czcRs of *Ralstonia eutropha*, czrRs of *Pseudomonas aeruginosa* and copRs of *Pseudomonas syringae*.

### 8.6.1  NONSPECIFIC LUX BIOREPORTERS

Nonspecific *lux* bioreporters are typically used for the detection of chemical toxins. They are usually designed to continuously bioluminesce. Upon exposure to a chemical toxin, either the cell dies or its metabolic activity is retarded, leading to a decrease in bioluminescent light levels.

Except this, UMT has also been used as a bioreporter for the selection of recombinant plasmids, as a marker for gene transcription in bacterial, yeast, and mammalian cells, and for the detection of toxic salts such as arsenite and antimonite.

Microbial bioreporters play an important role in environmental monitoring and ecotoxicology. Microorganisms that are genetically modified with reporter genes can be used in various formats to determine the bioavailability of chemicals and their effect on living organisms. Cyanobacteria are abundant in the photosynthetic biosphere and have considerable potential with regards to broadening bioreporter applications. Two recent studies described novel cyanobacterial reporters for the detection of environmental toxicants and iron availability [11].

The dose response relationship between seven commonly used herbicides and four luminescence-based bacterial biosensors was characterized. As herbicide concentration increased the light emitted by the test organism declined in a concentration dependent manner. These dose responses were used to compare the predicted vs. observed response of a biosensor in the presence of multiple contaminants. For the majority of herbicide interactions, the relationship was not additive but primarily antagonistic and sometimes synergistic. These biosensors provide a sensitive test and are able to screen a large volume and wide range of samples with relative rapidity and ease of interpretation. In this study biosensor technology has been successfully applied to interpret the interactive effects of herbicides in freshwater environments [12].

### 8.6.2  UV DETECTION

Microbial biosensors not only gives a significance response in pollutant food or Agriculture but can also report UV variation in ecosystem. *E. coli* strains containing plasmid borne fusions of the RecA promoter/operator region to the *Vibrio fischeri* lux gene were previously shown to increase their luminescence in the presence of DNA damage hazards. Lux genes also found in *Photorhabdus luminescense*, host may be *Salmonella typhi murium* instead of *E. coli,* use of *Salmonella* as host or *P. luminescence* reporter genes gives fastest response, however, *E. coli* shows high sensitivity because it has single copy of the *V. fischeri* lux fusion was integrated into the bacterial chromosomes [13].

Hydrocarbon degrading *Acinetobacter baumanni* strain S30, isolated from crude oil contaminated soil was inserted with a *lux* gene from the luciferase gene cassette *lux* CDABE. The modified bacteria helped to decide the level of soil contamination with crude oil [14].

## 8.7   FOOD AND AGRICULTUTRE

Immunodiagnostics and enzyme biosensors are two of the leading technologies that have a greatest impact on the food industry. The use of these two systems has reduced the time for detection of pathogens such as *Salmonella* to 24 hr and has provided detection of biological compounds such as cholesterol or chymotrypsin [15]. Biosensors analyzes Beta lactams in milk and presence of urea in milk which lead to production of "synthetic milk," the biocomponent part of the urea biosensor is an immobilized urease yielding bacterial cell biomass isolated from soil and is coupled to the ammonium ions selective electrodes of a potentiometric transducer. The membrane potential of all types of potentiometric cell based probes is related to the activity of electrochemically-detected product [16].

*Ralstonia eutropha* strain a whole-cell biosensor developed for the detection of bioavailable concentrations of $Ni_2+$ and $CO_2+$ in soil samples, the regulatory genes are fused to the bioluminescent lux CDABE reporter system. This modified bacteria used to study the accumulation of above said metals [17].

A biosensor for detecting the toxicity of polycylic aromatic hydrocarbons (PAHs) contaminated soil has been successfully constructed using an immobilized recombinant bioluminescent bacterium, GC2 (lac$^l$uxCDABE), which constitutively produces bioluminescence. This biosensor system using a biosurfactant may be applied as an in-situ biosensor to detect the toxicity of hydrophobic contaminants in soils and for performance evaluation of PAH degradation in soils [18].

A bioluminescent whole-cell reporter *Ralstonia eutropha* strain developed by using with a tfdRP(DII)-luxCDABE fusion gene for detection of weedicides like 2, 4-dichlorophenoxyacetic acid and 2,4-dichlorophenol in the contaminated soils [19].

A Synechococcus PglnA¹uxAB fusion for estimation of nitrogen bioavailability to freshwater cyanobacteria was done by fusing the promoter of the glutamine synthetase-encoding gene, P. glnA, from *Synechococcus* sp. strain PCC7942 to the luxAB luciferase-encoding genes of the bioluminescent bacterium *Vibrio harveyi* [20].

Tagging potato leaf roll virus with the jellyfish green fluorescent protein gene helped to glow the plant immediately after infection [21].

## 8.8   CONCLUSION

Biosensors represent a rapidly expanding field, at the present time, with an estimated 60% annual growth rate. Conventional electrochemical biosensors have been widely used in various fields of science. New concept of using genetically engineered microbial biosensors (bioreproters) are one of the most expanding field because it is highly efficient by emitting easily detectable, long duration signals. Recently developed bioreporter technology providing a robust, cost-effective, quantitative method for rapid and selective detection and monitoring of physical and chemical and biological agents in applications as far ranging as medical diagnostics, precision agriculture, environmental monitoring, food safety, process monitoring & control, space and research, defense, etc.

Bioreporters attractiveness lies in the fact that they can often be implemented in real-time, on-line bioassays within intact, living cell systems. Since molecular biology and microbiology are growing with rapid pace in near future we can able to see new and wide rage of genetically modified microbes as biosensors.

## KEYWORDS

- Acinetobacter baumanni
- Aequorea victoria
- Photorhabdus luminescense
- Pseudomonas fluorescence
- Pseudomonas putida
- Renilla reniformis
- Saccharomyces cerevisiae
- Salmonella typhi murium
- Vibrio fischeri

## REFERENCES

1. Mark Frauenfelder, Biosniffing Wireless Data, January **2003**.
2. Cancer Hirschmann, F.; Verhoeyen, E.; Wirt,h D.; Bauwens, S.; Hauser, H.; Rudert, M. Vital marking of articular chondrocytes by retroviral infection using green fluorescence protein, Osteoarthritis Cartilage.; February **2002**, *2*, 109–118.
3. Choy, G.; Choyke, P.; Libutti, S. K.; Current advances in molecular imaging: noninvasive in vivo bioluminescent and fluorescent optical imaging in cancer research, *Mol Imaging.;* October **2003**, *2(4)*, 303–312.
4. Choy, G.; O'Connor, S.; Diehn, F. E.; Costouros, N.; Alexander, H. R.; Choyke, P.; Libutti, S. K.; Comparison of noninvasive fluorescent and bioluminescent small animal optical imaging, *Biotechniques.;* November **2003**, *35(5)*, 1022–6, 1028–1030.
5. Bachmann, A. S.; The role of polyamines in human cancer: prospects for drug combinationthera[pies, *Hawaii Med J.;* December **2004**, *63(12)*, 371–374.
6. Wolkwicz, R.; Nolna, G. P.; Gene therapy progress and prospects: noval gene therapy approches for AIDS, gene therapy, February **2005**, [Epud ahead of print].
7. Piunno, P. A.; Krull, U. J.; Trends in the development of nucleic acid biosensors for medical diagnostics. *Anal Bioanal Chem.;* February **2005**, [Epub ahead of print]
8. David A.; Relman, M.D. Shedding Light on Microbial Detection, *New J. Med.*; November **2003**, *349*, 2162–2163.
9. Babich, H.; Devanas, M. A.; Stotzky G, The mediation of mutagenicity and clastogenicity of heavy metals by physicochemical factors, *Environ Res.*; August **1985**, *37(2)*, 253–286.
10. Ren, S.; Frymier, P. D. Toxicity of metals and organic chemicals evaluated with bioluminescence assays, *Chemosphere.*; February **2005**, *58(5)*, 543–550.
11. Bachmann, T. Transforming cyanobacteria into bioreporters of biological relevance, *Trends Biotechnol.*; June **2003**, *21(6)*, 247–249.

12. Strachan, G.; Preston, S; Maciel, H.; Porter, A. J.; Paton, G. I.; Use of bacterial biosensors to interpret the toxicity and mixture toxicity of herbicides in freshwater, *Water Res.*; October **2001**, *35(14)*, 3490–3495.
13. Rosen, R.; Davidov, LaRossa, R.A.; Belkin, S.; Microbial sensors of ultraviolet radiation based on rec'ux gene fusion, *Appl Biochem Biotechnol*, November–December **2000**, *89(2–3);* 151–160.
14. Mishra S.; Sarma, P. M.; Lal B.; Crude oil degradation efficiency of a recombinant Acinetobacter baumannii strain and its survival in crude oil-contaminated soil microcosm. *FEMS Microbiol Lett.;* June **2004**, *235(2)*, 323–331.
15. Richter, E. R.; Biosensors: Applications for dairy food industry. *J Dairy Sci.;* October **1993**, *76(10)*, 3114–3117.
16. Verma, N.; Singh, M.; A disposable microbial based biosensor for quality control in milk. *Biosens Bioelectron.;* September **2003**, *18(10)*, 1219–1224.
17. Tibazarwa, C.; Corbisier, P.; Mench, M.; Bossus, A.; Solda, P.; Mergeay, M.; Wyns, L.; van der Lelie, D.; A microbial biosensor to predict bioavailable nickel in soil and its transfer to plants, *Environ Pollut.;* **2001**, *113(1)*, 19–26 [ni].
18. Gu, M.B.; Chang, S. T.; Soil biosensor for the detection of PAH toxicity using an immobilized recombinant bacterium and a biosurfactant, *Biosens Bioelectron.;* December **2001**, *16(9–12)*, 667–674 [pah].
19. Hay, A.G; Rice, J. F.; Applegate, B. M.; Bright, N. G.; Sayler, G. S.; A bioluminescent whole-cell reporter for detection of 2, 4-dichlorophenoxyacetic acid and 2, 4-dichlorophenol in soil. *Appl Environ Microbiol.;* October **2000**, *66(10)*, 4589–4594.
20. Gillor, O.; Harush, A; Hada,s O.; Post, A. F.; Belkin S, A Synechococcus PglnA'uxAB fusion for estimation of nitrogen bioavailability to freshwater cyanobacteria, *Appl Environ Microbiol.;* March **2003**, *69(3)*, 1465–1474.
21. Nurkiyanova, K. M.; Ryabov, E. V.; Commandeur U, Duncan, G. H.; Canto T, Gray, S. M.; Mayo, M. A.; Taliansky, M. E.; Tagging potato leafroll virus with the jellyfish green fluorescent protein gene, *J Gen Virol.;* March **2003**, *81(Pt 3)*, 617–626.

# CHAPTER 9

# ON DEVELOPMENT OF SOLAR CLOTH BY ELECTROSPINNING TECHNIQUE

VAHID MOTTAGHITALAB

## CONTENTS

## 9.1   INTRODUCTION

In recent years, renewable energies attract considerable attention due to the inevitable end of fossil fuels and due to global warming and other environmental problems. Photovoltaic solar energy is being widely studied as one of the renewable energy sources with key significance potentials and a real alternate to fossil fuels. Solar cells are in general packed between w80, brittle and rigid glass plates. Therefore, increasing attention is being paid to the construction of lighter, portable, robust, multipurpose and flexible substrates for solar cells. Textiles substrates are fabricated by a wide variety of processes, such as weaving, knitting, braiding and felting. These fabrication techniques offer enormous versatility for allowing a fabric to conform to even complex shapes. Textile fabrics not only can be rolled up for storage and then unrolled on site but also they can also be readily installed into structures with complex geometries.

Textiles are engaging as flexible substrates in that they have a enormous variety of uses, ranging from clothing and household articles to highly sophisticated technical applications. Last innovations on photovoltaic technology have allowed obtaining flexible solar cells, which offer a wide range of possibilities, mainly in wearable applications that need independent systems. Nowadays, entertainment, voice and data communication, health monitoring, emergency, and surveillance functions, all of which rely on wireless protocols and services and sustainable energy supply in order to overcome the urgent needs to regular battery with finite power. Because of their steadily decreasing power demand, many portable devices can harvest enough energy from clothing-integrated solar modules with a maximum installed power of 1–5 W. [1]

Increasingly textile architecture is becoming progressively of a feature as permanent or semipermanent constructions. Tents, such as those used by the military and campers, are the best known textile constructions, as are sun shelter, but currently big textile constructions are used extensively for exhibition halls, sports complexes and leisure and recreation centers. Although all these structures provide protection from the weather, including exposure to the sun, but solar concept offers an additional precious use for providing power. Many of these large textile architectural constructions cover huge areas, sufficient to supply several kilowatts of power.

Even the fabric used to construct a small tent is enough to provide a few hundred watts. In addition to textile architecture, panels made from robust solar textile fabrics could be positioned on the roofs of existing buildings. Compared to conventional and improper solar panels for roof structures lightweight and flexible solar textile panels is able to tolerate load-bearing weight without shattering.

Moreover, natural disaster extensively introduces the huge potential needs the formulation of unusual energy package based on natural source. Over the past five years, more than 13 million *people have lost their home* and possessions because of earthquake, bush fire, flooding or other natural disaster. The victims of these disasters are commonly housed in tents until they are able to rebuild their homes. Whether they stay in tented accommodation for a short or long time, tents constructed from solar textile fabrics could provide a source of much needed power. This power could be stored in daytime and used at night, when the outdoor temperature can often fall. There are also a number of other important potential applications. The military would benefit from tents and field hospitals, especially those in remote areas, where electricity could be generated as soon as the structure is assembled.

## 9.2   THE BASIC CONCEPT OF SOLAR CELL

In 1839 French scientist, Edmond Becquerel found out photovoltaic effect when he observed increasing of electricity generation while light exposure to the two metal electrodes immersed in electrolytic solution [2]. Light is composed of energy packages known as photons. Typically, when a matter exposed to the light, electrons are excited to a higher level within material, but they return to their initial state quickly. When electrons take sufficient energy more than a certain threshold (band gap), move from the valance band to the conduction band holes with positive charge will be created. In the photovoltaic effect electron-hole pairs are separated and excited electrons are pulled and fed to an external circuit to buildup electricity [3] (Fig. 1).

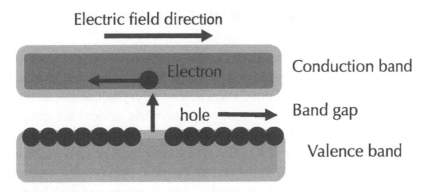

**FIGURE 1** Electron excitation from valence band to conduction band.

An effective solar cell generally comprises an opaque material that absorbs the incoming light, an electric field that arises from the difference in composition between the semiconducting layers comprising the absorber, and two electrodes to carry the positive and negative charges to the electrical load. Designs of solar cells differ in detail but all must include the above features (Fig. 2).

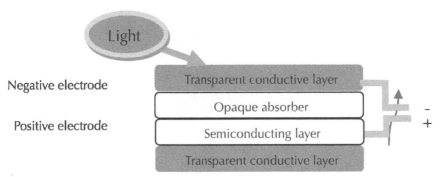

**FIGURE 2** The general structure of a Solar cell.

Generally, Solar cells categorized into three main groups consisting of inorganic, organic and hybrid solar cells (Dye sensitized solar cells). Inorganic solar cells cover more than 95% of commercial products in solar cells industry.

## 9.2.1  ELECTRICAL MEASUREMENT BACKGROUND

All current–voltage characteristics of the photovoltaic devices were measured with a source measure unit in the dark and under simulated solar simulator source was calibrated using a standard crystalline silicon diode. The current-voltage characteristics of Photovoltaic devices are generally characterized by the short-circuit current $(I_{sc})$, the open-circuit voltage $(V_{oc})$, and the fill factor $(FF)$. The photovoltaic power conversion efficiency (η) of a solar cell is defined as the ratio between the maximum electrical power $(P_{max})$ and the incident optical power and is determined by Eq. (1) [4].

$$\eta = \frac{I_{sc} \times V_{oc} \times FF}{P_{in}} \quad (1)$$

In Eq. (1), the short-circuit current $(I_{sc})$ is the maximum current that can run through the cell. The open circuit voltage $(V_{oc})$ depends on the highest occupied molecular orbital (homo)level of the donor (p-type semiconductor quasi Fermi level) and the lowest unoccupied molecular orbital(lumo) level of the acceptor (n-type semiconductor quasi Fermi level), linearly. $P_{in}$ is the incident light power density. $FF$, the fill-factor, is calculated by dividing $P_{max}$ by the multiplication of $I_{sc}$ and $V_{oc}$ and this can be explained by the following Eq. (2):

$$FF = \frac{I_{mpp} \times V_{mpp}}{I_{sc} \times V_{oc}} \quad (2)$$

In the Eq. (2), $FF_{mpp}$ and $I_{mpp}$ represent, respectively, the voltage and the current at the maximum power point (MPP), where the product of the voltage and current is maximized [4].

## 9.2.2  INORGANIC SOLAR CELL

Inorganic solar cells based on semiconducting layer architecture can be divided into four main categories including P-N homo junction, hetrojunction either P-I-N or N-I-P and multi junction.

The P-N homojunction is the basis of inorganic solar cells in which two different doped semiconductors (n-type and p-type) are in contact to make solar cells (Fig. 3a). P-type semiconductors are atoms and compounds with fewer electrons in their outer shell, which could create holes for the electrons within the lattice of p-type semiconductor. Unlike p-types semiconductors, the n-type have more electrons in their outer shell and sometimes there are exceed amount of electron on n-type lattice result lots of negative charges [5].

Compared to homojunction structure amorphous silicon thin-film cells use a P-I-N hetrojunction structure, whereas cadmium telluride (CdTe) cells use a N-I-P arrangement. The overall picture embrace a three-layer sandwich with a middle intrinsic (i-type or undoped) layer between an N-type layer and a P-type layer. (Fig. 3b). Multiple junction cells have several different semiconductor layers stacked together to absorb different wavebands in a range of spectrum, producing a greater voltage per cell than from a single junction cell which most of the solar spectrum to electricity lies in the red (Fig. 3c). Variety of semiconducting material such as single and poly crystal silicon, amorphous silicon, CdTe, Copper Indium/Gallium Di Selenide (CIGS) have been employed to form inorganic solar cell based on layers configuration to enhance absorption efficiency, conversion efficiency, production and maintenance cost.

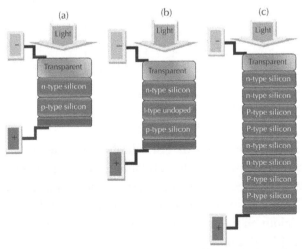

FIGURE 3 A general scheme of inorganic solar cell (a) P-N homojunction (b) P-I-N hetrojunction (c) multijunction.

### 9.2.3   ORGANIC  SOLAR  CELLS  (OSCS)

Photoconversion mechanism in organic or excitonic solar cells is differing from conventional inorganic solar cells in which exited mobile state are made by light absorption in electron donor. However, light absorption creates free electron-hole pairs in inorganic solar cells [5]. It is due to law dielectric constant of organic materials and weak noncovalent interaction between organic molecules. Consequently, exciton dissociation of electron-hole pairs occurs at the interface between electron donor and electron acceptor components [6]. Electron donor and acceptor act as semiconductor p-n junction in inorganic solar cells and should be blended together to prevent electron-hole recombination (Fig. 4).

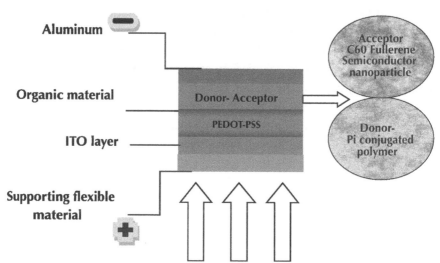

**FIGURE 4**   Schematic of organic solar cell.

There are two main types of PSCs including: bilayer heterojunction and bulk-heterojunction [7]. Bulk-heterojunction PSCs are more attractive due to their high surface area junction that increases conversion efficiency. This type of polymer solar cell consists of Glass, ITO, PEDOT: PSS, active layer, calcium and aluminum in which conjugated polymer are used as active layer [8]. The organic solar cells with maximum conversion efficiency about 6% still are at the beginning of development and have a long way to go to compete with inorganic solar cells. Indeed, the advantages of polymers including low-cost deposition in large areas, low weight on flexible substrates and sufficient efficiency are promising advent of new type of solar cells [9]. Conjugated small molecule attracted as an alternative approach of organic solar cells. Development of small molecule for OSCs interested because of their properties such as well-defined molecular structure, definite molecular weight, high purity, easy purification, easy mass-scale production, and good batch-to-batch reproducibility [10–12].

## 9.2.4  DYE-SENSITIZED SOLAR CELLS

Dye-sensitized solar cells (DSSC) use a variety of photosensitive dyes and common, flexible materials that can be incorporated into architectural elements such as window panes, building paints, or textiles. DSSC technology mimic photosynthesis process whereby the leaf structure is replaced by a porous titania nanostructure, and the chlorophyll is replaced by a long-life dye. The general scheme of DSSC process is shown in Fig. 5. Although traditional silicon-based photovoltaic solar cells currently have higher solar energy conversion ratios, dye-sensitive solar cells have higher overall power collection potential due to low-cost operability under a wider range of light and temperature conditions, and flexible application [13]. Oxide semiconductors materials such as $TiO_2$, $ZnO_2$ and $SnO_2$ have a relatively wide band gap and cannot absorb sunlight in visible region and create electron. Nevertheless, in sensitization process, visible light could be absorbed by photosensitizer organic dye results creation of electron. Consequently, excited electrons are penetrated into the semiconductor conduction band. Generally, DSSC structures consist of a photoelectrode, photosensitizer dye, a redox electrolyte, and a counter electrode. Photoelectrodes could be made of materials such as metal oxide semiconductors. Indeed, oxide semiconductor materials, particularly $TiO_2$, are choosing due to their good chemical stability under visible irradiation, nontoxicity and cheapness. Typically, $TiO_2$ thin film photoelectrode prepared via coating the colloidal solution or paste of $TiO_2$ and then sintering at 450°C to 500°C on the surface of substrate, which led to increase of dye absorption drastically by $TiO_2$ [14]. The substrate must have high transparency and low ohmic resistance to high performance of cell could be achieved. Recently many researches focused on the both organic and inorganic dyes as sensitizer regarding to their extinction coefficients and performance. Among them, B4 (N3): ruL2(NCS)2:L=(2,2'-bipyridyl-4,4'-dicarboxylic acid) and B2(N719): {cis-bis (thiocyanato)-bis(2,20-bipyridyl-4,40-dicarboxylato)-ruthenium(II) bis-tetrabutylammonium} due to its outstanding performance was interested.(Scheme 1). Oxide semiconductors materials such as $TiO_2$, $ZnO_2$ and $SnO_2$ have a relatively wide band gap and cannot absorb sun light in visible region and create electron. Nevertheless, in sensitization process, visible light could be absorbed by photosensitizer organic dye results creation of electron.

**FIGURE 5** Schematic of Dye Senetesized Solar Cell (DSSC)

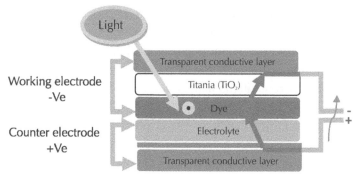

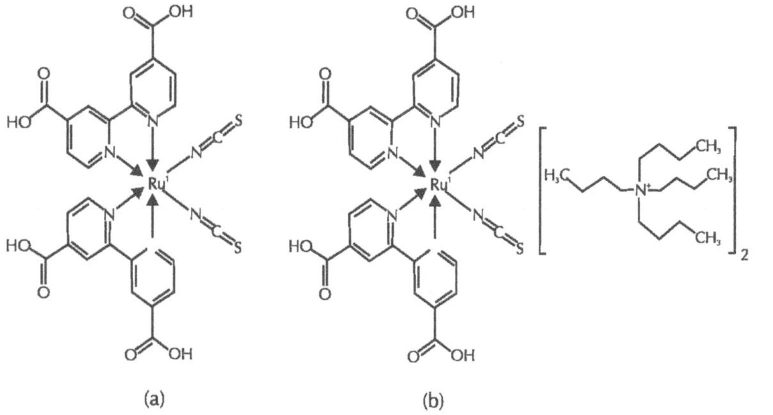

(a)                                                    (b)

**SCHEME 1** The chemical structure of (a) B4 (N3): ruL2(NCS)2 L(2,2'-bipyridyl-4,4'-dicarboxylic acid), (b) B2(N719) {cis-bis (thiocyanato)-bis(2,20-bipyridyl-4,40-dicarboxylato)-ruthenium(II) bis-tetrabutylammonium.

Consequently, excited electrons are penetrated into the semiconductor conduction band. Generally, DSSC structures consist of a photoelectrode, photosensitizer dye, a redox electrolyte, and a counter electrode. Photoelectrodes could be made of materials such as metal oxide semiconductors.

Indeed, oxide semiconductor materials, particularly $TiO_2$, are choosing due to their good chemical stability under visible irradiation, nontoxicity and cheapness. Typically, $TiO_2$ thin film photoelectrode prepared via coating the colloidal solution or paste of $TiO_2$ and then sintering at 450∘C to 500∘C on the surface of substrate, which led to increase of dye absorption drastically by $TiO_2$ [14]. The substrate must have high transparency and low ohmic resistance to high performance of cell could be achieved. Recently many researches focused on the both organic and inorganic dyes as sensitizer regarding to their extinction coefficients and performance. Among them, B4 (N3): ruL2(NCS)2:L=(2,2′-bipyridyl-4,4′-dicarboxylic acid) and B2(N719): {cis-bis (thiocyanato)-bis(2,20-bipyridyl-4,40-dicarboxylato)-ruthenium(II) bis-tetrabutylammonium} due to its outstanding performance was interested (Scheme 1).

B2 is the most common high performance dye and a modified form of B4 to increase cell voltage. Up to now different methods have been performed to develop of new dyes with high molar extinction coefficients in the visible and near-IR in order to outperform N719 as sensitizers in a DSSC [15]. DNH2 is a hydrophobic dye, which very efficiently sensitizes wide band-gap oxide semiconductors, like titanium dioxide. DBL (otherwise known as "black dye") is designed for the widest range spectral sensitization of wide band-gap oxide semiconductors, like titanium dioxide up to wavelengths beyond 800 nm (Scheme 2).

In order to continuous electron movement through the cell, the oxidized dye should be reduced by electron replacement. The role of redox electrolyte in the DSSCs is to mediate electrons between the photoelectrode and the counter electrode. Common electrolyte used in the DSSC is based on $I^-/I_3^-$ redox ions [16]. The mechanism of photon to current has been summarized in the following equations:

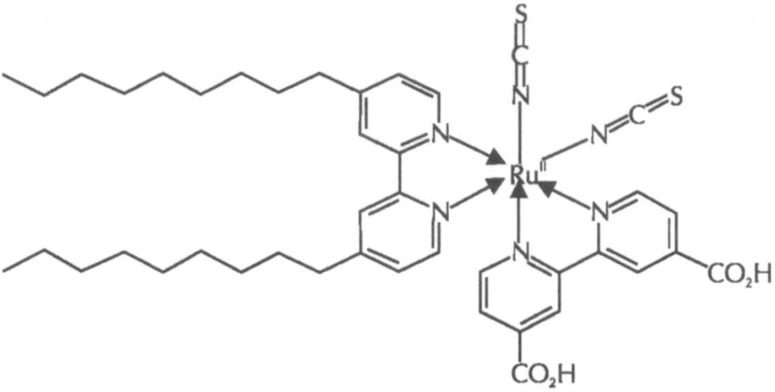

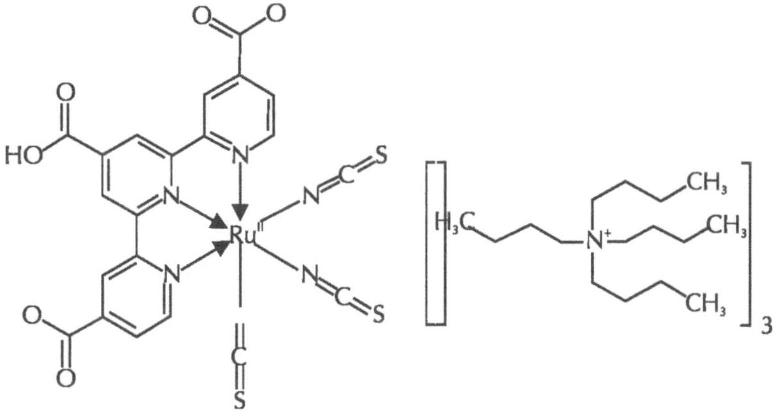

**SCHEME 2**   The chemical structure of (a) DNH2 (Z907) RuLL'(NCS)2, L=2,2'-bipyridyl-4,4'-dicarboxylic acid, L'= 4,4'-dinonyl-2,2'-bipyridine (b) DBL (N749) [RuL(NCS)3]: 3 TBA L= 2,2':6,'2"-terpyridyl-4,4,'4"-tricarboxylic acid TBA=tetra-n-butylammonium.

$$\text{Dye} + \text{light} \rightarrow \text{Dye}^* + e^- \qquad\qquad\qquad\qquad (a)$$

$$\text{Dye}^* + e^- + \text{TiO}_2 \rightarrow e^-(\text{TiO2}) + \text{oxidized Dye} \qquad (b)$$

$$\text{Oxidized Dye} + 3/2\text{I}^- \rightarrow \text{Dye} + 1/2\text{I}^{-3} \qquad\qquad (c)$$

$$1/2\text{I}^-_3 + e^-(\text{counter electrode}) \rightarrow 3/2\text{I}^- \qquad\qquad (d)$$

Despite of many advantages of DSSCs, still lower efficiency compared to commercialized inorganic solar cells is a challenging area. Recently, one-dimensional nanomaterials, such as nanorods, nanotubes and nanofibers, have been proposed to replace the nanoparticles in DSSCs because of their ability to improve the electron transport leading to enhanced electron collection efficiencies in DSSCs.

Subsequent sections attempts to provide fundamental knowledge to general concept of textile solar cells and their recent progress based on. Of particular interest are electrospun $TiO_2$ nanofibers playing the role as a key material in DSSCs and other organic solar cell, which have been shown to improve the electron transport efficiency and to enhance the light harvesting efficiency by scattering more light in the red part of the solar spectrum. A detailed review on cell material selection and their effect on energy conversion are considered to elucidate the potential role of nanofiber in energy conversion for textile solar cell applications.

## 9.3   TEXTILE SOLAR CELLS

Clothing materials either for general or specific use are passive and the ability to integrate electronics into textiles provides great opportunity as smart textiles to achieve revolutionary improvements in performance and the realization of capabilities never before imagined on daily life or special circumstances such as battlefield. In general, smart textiles address diverse function to withstand an interactive wearable system. Development, incorporation and interconnection of flexible electronic devices including sensor, actuator, data processing, communication, internal network and energy supply beside basic garment specifications sketch the road map toward smart textile architecture. Regardless of the subsystem functions, energy supply and storage play a critical role to propel the individual functions in overall smart textile systems.

The integration of photovoltaic (PVs) into garments emerges new prospect of having a strictly mobile and versatile source of energy in communications equipment, monitoring, sensing and actuating systems. Despite of extremely good power efficiency, most conventional crystalline silicon based semiconductor PVs are intrinsically stiff and incompatible with the function of textiles where flexibility is essential. Extensive research has been conducted to introduce the novel potential candidates for shaping the textile solar cell (TSC) puzzle. In particular, polymer-based organic solar cell materials have the advantages of low price and ease of operation in comparison with silicon-based solar cells. Organic semiconductors, such as conductive polymers, dyes, pigments, and liquid crystals, can be manufactured cheaply and used in organic solar cell constructions easily. In the manufacturing process of organic solar cells, thin films are prepared using specific techniques, such as vacuum evaporation, solution processing, printing [17, 18], or nanofiber formation [19] and electrospinning [20] at room temperatures. Dipping, spin coating, doctor blading, and printing techniques are mostly used for manufacturing organic solar cells based on conjugated polymers [17]. Recent TSC studies revealed two distinctive strategies for developing flexible textile solar cell and its sophisticated integration.

1. The first strategy involved the simple incorporation of a polymer PV on a flexible substrate Such as poly thyleneterphthalate (PET) directly into the clothing as a structural power source element.

2. The second strategy was more complicated and involved the lamination of a thin anti reflective layer onto a suitably transparent textile material followed by plasma, thermal or chemical treatment. The next successive step focuses on application of a photoanode electrode onto the textile material. Subsequent procedure led to the deposition of the active material and finally evaporation of the cathode electrode complete the device as a textile PV composed of organic, inorganic and also their composites.

Regardless of many gaps need to be bridged before large-scale application of this technology, the TSC fabrication based on second strategy may be envisaged through two routes to solve pertinent issues of efficiency and stability. The solar cell architecture in first approach is founded based on knitted or woven textile substrate, however, second alternative follows a roadmap to develop a wholly PV fiber for further knitting or weaving process that may form energy-harvesting textile structures in any shape and structure.

Irrespective to fabric or fiber shaped of the photovoltaic unit, the light penetration and scattering in photo anode layer needs a waveguide layer.

This basic requirement naturally mimicked by polar bear hair. The optical functions of the polar bear hair are scattering of incident light into the hair, luminescence wave shift and wave-guide properties due to total reflection. The hair has an opaque, rough-surfaced core, called the medulla, which scatters incident light. The simulated synthetic coreshell fiber can be manufactured through spinning of a core fiber or with sufficient wave-guiding properties followed by finishing with an optically active, i.e., fluorescing, coating as a shell to achieve a polar bear hair' effect [21]. As described in Tributsch's original work [22], a high-energy conversion can be expected from high frequency shifts as difference between the frequency of absorbed and emitted light. According to Tributsch et al. [22] for the polar bear hair, the frequency shift is of the order of $2 \times 10^{14}$ Hz.

In principle, the refractive index varies over the fiber diameter and this would provide a certain wave-guiding property of the fiber. In charge of wave-guiding property most specifically With regard to manufacturing on a larger scale, fiber morphology, crystalinity, alignment, diameter and geometry can be altered for preferred optical performance as solar energy transducer. The focus on both approached of second strategy and illuminate their opportunities and challenges is main area of interest in next parts.

### 9.3.1  RECENT PROGRESS OF PV FIBER

Fiber shaped organic solar cells has been subject of a few patent, project and research papers. Polymers, small molecules and their combinations were used as light-absorbing layers in previous studies. A range of synthetic substrate with various level of flexibility including optical [23], polyimide [24] and poly propylene (PP) [25] subjected to processes in which functional electrodes and light absorbing layer continuously forms on fiber scaffold. In recent PV fiber studies, conducting polymers such as P3HT in combination to small molecules nanostructure materials such as branched fullerene plays main role as photoactive material. For instance, Ref. [23] introduce a light absorbing layer on optical fiber composed of poly (3-hexylthiophene) (P3HT): phenyl-C61-butyric acid methyl ester (P3HT:PCBM) (Fig. 6). However, the light was traveling through the optical fiber and generating hole–electron pairs, the 100 nm top metal electrode (which does not let the light transmit from outside) was used to collect the electrons [23].

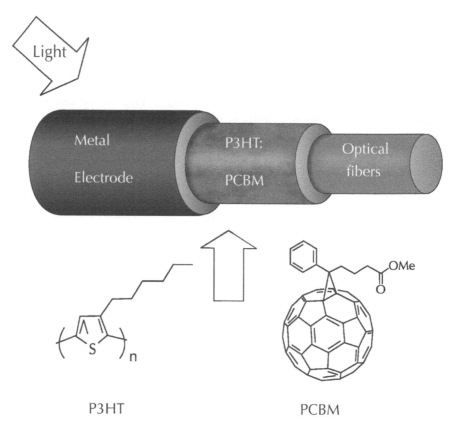

**FIGURE 6**   Simplified pattern of heterojunction photoactive layer in fiber solar cell.

One of the important challenges of flexible solar cell concentrated in hole collecting electrode, which is most widely, used ITO as a transparent conducting material. However, the inclusion of ITO layer in flexible solar cell could not be applicable. The restrictions are mostly due to the low availability and expense of indium, employment of expensive vacuum deposition techniques and providing high temperatures to guarantee highly conductive transparent layers. Accordingly, there are some ITO-free alternative approaches, such as using carbon nanotube (CNT) layers or different kinds of poly(3,4-ethylenedioxythiophene):poly(stirenesulfon ate) (PEDOT:PSS) and its mixtures [26–28], or using a metallic layer [29] to perform as a hole-collecting electrode. (Scheme 3)

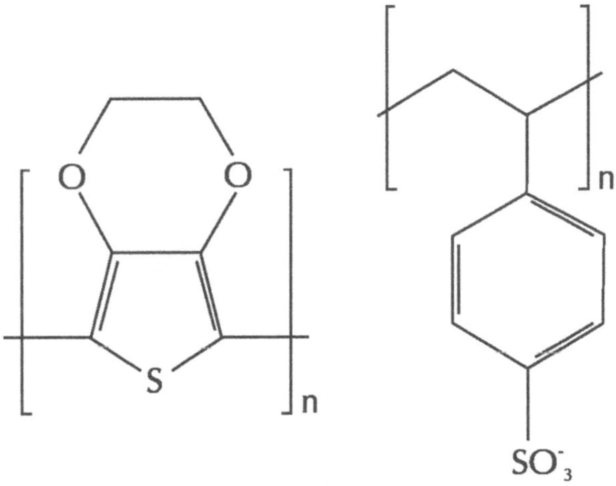

**SCHEME 3**    The chemical structure of poly (3,4-ethylenedioxythiophene):poly(stirenes ulfonate).

The ITO free hole collecting layer was realized using highly conductive solution of PEDOT:PSS as a polymer anode that is more convenient for textile substrates in terms of flexibility, material cost, and fabrication processes compared with ITO material. Based on procedure described in Ref. [25] a sophisticated and simple design was presented to show how thin and flexible could be a solar cell panel (Fig. 7).

Based on implemented pattern, the sunbeams entered into the photoactive layer with 4–10 mm² active area by passing through a 10 nm of lithium fluoride/aluminum (LiF/Al) layer as semitransparent cathode outer electrode.

Table 1 gives the current density data versus voltage characteristics of the photovoltaic fibers consisting of P3HT: PCBM and MDMO-PPV: PCBM blends. Based on given results of open-circuit voltage, short-circuit current density and fill factor for two types of photoactive material, and also Eqs. (1) and (2) in Section 2.2, the power conversion efficiency of the MDMO-PPV: PCBM based photovoltaic fiber was higher than the P3HT: PCBM based photovoltaic fiber.

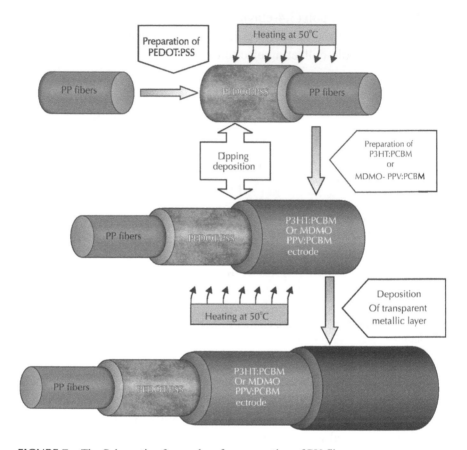

**FIGURE 7**   The Schematic of procedure for preparation of PV fiber.

**TABLE 1**   Photoelectrical characteristics of photovoltaic fibers having different having different photoactive layers (P3HT: PCBM and MDMO-PPV:PCBM) [25].

| Solar cell pattern | $V_{oc}$(mV) | $I_{sc}$(mA/cm$^2$) | FF (%) | $\eta$ (%) |
|---|---|---|---|---|
| **PP\|PEDOT:PSS\| P3HT:PCBM\|LiF/Al** | 360 | 0.11 | 24.5 | 0.010 |
| **PP\|PEDOT:PSS\|   MDMO-PPV:PCBM\| LiF/Al** | 300 | 0.27 | 26 | 0.021 |
| **ITO\|PEDOT:PSS\|  MDMO-PPV:PCBM\| LiF/Al** | 740 | 4.56 | 43.4 | 1.46 |

Comparing the solar cell characteristics of second and third pattern shows greater performance of ITO|PEDOT:PSS| MDMO-PPV:PCBM| LiF/Al rigid organic solar cell compared to PP|PEDOT:PSS| MDMO-PPV:PCBM| LiF/Al fiber solar cell. Since a same cathode, anode and photoactive material used in both pattern, the higher power conversion efficiency of ITO solar cell can be attributed to different wave-guide property and transparency of cell pattern in sun's ray entrance angle (i.e., LiF/AL versus ITO glass). Since using ITO is strictly restricted for PV fiber, enhancing the power conversion efficiency of needs to improving existing materials and techniques. In particular, the optical band gap of the polymers used as the active layer in organic solar cells is very important. Generally, the best bulk heterojunction devices based on widely studied P3HT: PCBM materials are active for wavelengths between 350 and 650 nm. Polymers with narrow band gaps can absorb more light at longer wavelengths, such as infra-red or near-infra-red, and consequently enhance the device efficiency. Low band gap polymers (<1.8 eV) can be an alternative for better power efficiency in the future, if they are sufficiently flexible and efficient for textile applications [30, 31]. The variety of factors influence on polymer band gaps, which can be categorized as intrachain charge transfer, substituent effect, $\pi$-conjugation length. Systematically the fused ring low band gap copolymer composes of a low energy level electron acceptor unit coupled with a high energy level electron donor unit. The band gap of the donor/acceptor copolymer is determined by the HOMO of the donor and LUMO of the acceptor, and therefore, a high energy level of the HOMO of the donor and a low energy level of the LUMO of the acceptor results in a low band gap [32].

The substituent on the donor and acceptor units can affect the band gap. The energy level of the HOMO of the donor can be enhanced by attaching electron-donating groups (EDG), such as thiophene and pyrrole. Similarly, the energy level of the LUMO of the acceptor is lowered, when electron-withdrawing groups (EWG), such as nitrile, thiadiazole and pyrazine, are attached. This will result in improved donor and acceptor units, and hence, the band gap of the polymer is decreased.

## 9.3.2   PV INTEGRATION IN TEXTILE

Having a complete functional textile solar cell motivates the researchers to attempt an approach for direct incorporation of photovoltaic cell elements onto the textile. The textile substrates inherently scatter most part of the incident light outward. Therefore, it was found necessary to apply a layer of the very flexible polymer PE onto the textile substrate to have a surface compatible with a layered device. The textile-PE substrate was plasma treated before application of the transparent PEDOT electrode in order to obtain good adhesion of the PEDOT layer to the PE carrier. Then screen-printing was employed for the application of the active polymer poly[2-methoxy-5-(2'-ethylhexyloxy)-p-phenylene vinylene] (MEH-PPV) [33].

The traditional solar cell geometry was reinvented in fractal forms that allow the building of structured modules by sewing the 25–40 cm cells realized. Figure 8 shows a step-by-step approach for fabrication textile solar cell pattern based on polymer photo absorbing layer.

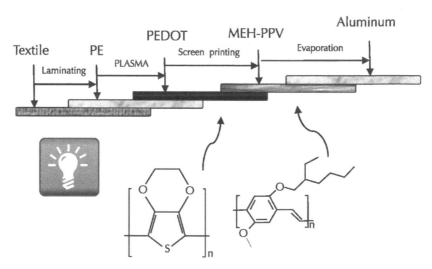

**FIGURE 8**   A typical fabrication procedure and key elements of textile solar cell.

The pattern designed was particularly challenging for application in solar cells and reduced the active area to 190 cm$^2$ (19% of the real area). The best module output power was found to be 0.27 mV with a Isc = 3:8 μA, Voc = 275 mV and a FF% of 25.7%. The pattern designed allows connections in different site of the cloth cell with reproducible performances within 5–10%.

## 9.4  NANOFIBERS AS A POTENTIAL KEY ELEMENT IN TEXTILE SOLAR CELLS

Previous sections present variety of solar cell structure and their corresponding elements and power conversion performance to indicate opportunities and challenges of producing of solar energy harvesting module based on a wholly flexible textile based photovoltaic unit. Current state of Textile Solar Cells is extremely far from commercial inorganic heterojunction solar cells that showing around 45% conversion efficiency. Current section addresses promising potential of nanofiber 1D morphology to be used as solar cell elements. Of particular, enhancement of photovoltaic unit demanding properties is a great of importance. Two different strategies can be presumed including integration of functional photoanode, photo cathode, scattering layer, photoactive or acceptor – donor materials in the form of nanofiber on to textile substrate or developing fully integrated multilayer nonwoven solar cloth.

### 9.4.1  ELECTROSPUN NANOFIBER

Fibers with a diameter of around 100 nm are generally classified as nanofibers. What makes nanofibers of great interest is their extremely small size. Nanofibers compared to conventional fibers, with higher surface area to volume ratios and smaller pore size, offer an opportunity for use in a wide variety of applications. To date, the most successful method of producing nanofibers is through the process of electrospinning. The electrospinning process uses high voltage to create an electric field between a droplet of polymer solution at the tip of a needle and a collector plate. When the electrostatic force overcomes the surface tension of the drop, a charged, continuous jet of polymer solution is ejected. As the solution moves away from the needle and toward the collector, the solvent evaporates and jet rapidly thins and dries. On the surface of the collector, a nonwoven web of randomly oriented solid nanofibers is deposited. Material properties such as melting temperature and glass transition temperature as well as structural characteristics of nanofiber webs such as fiber diameter distribution, pore size distribution and fiber orientation distribution determine the physical and mechanical properties of the webs. The surface of electrospun fibers is important when considering end-use applications. For example, the ability to introduce porous surface features of a known size is required if nanoparticles need to be deposited on the surface of the fiber.

The conventional setup for producing a nonwoven layer can be manipulated to fabricate diverse profile and morphology including oriented [34], Core shell [35] and hollow [36] nanofiber. Figure 9 shows the latest nanofiber profiles and its corresponding electrospinning production instrument. The variety and propagation of nanofiber products opens new horizon for development of functional profile respect to demanding application. Amongst developed techniques, coaxial electrospinning forms coreshell and/or hollow nanofiber through combination of different materials in the core or shell side, novel properties and functionalities for nanoscale devices can be found.

(a)

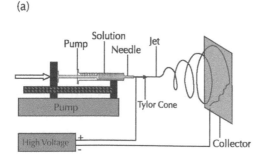 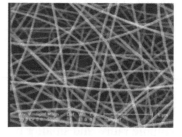

FIGURE 9    *(Continued)*

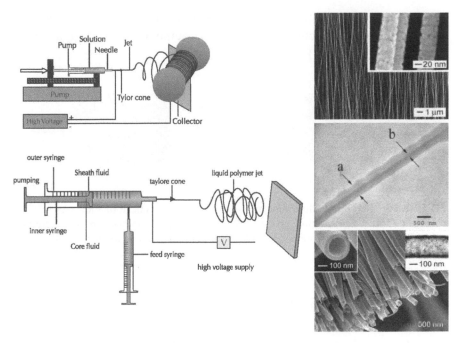

**FIGURE 9** Electrospinning setup and its corresponding nanofiber profile (a) conventional nanofiber (b) Oriented nanofiber [34] (c) Core shell nanofiber [35], (d) Hollow nanofiber [ 36].

Increasing demands for the manufacturing of bi-component structures, in which one is surrounded by the other or the particles of one are encapsulated in the matrix of the other, at the micro-or nano-level, show potential for a wide range of uses. Application includes minimizing chances of decomposition of an unstable material, control releasing a substance to a particular receptor and improving mechanical properties of a core polymer by its reinforcing with another material. The electrospinneret consists of concentric inner and outer syringe by witch two fluids are introduced to the spinneret, one in the core of the inner syringe and the other in the space between in the inner syringe and outer syringe. The droplet of the sheath solution elongates and stretches due to the repulsing between charges and form a conical shape. When the applied voltage increases, the charge accumulation reaches a certain value so a thin jet extends from the cone.

The stresses are generated in the sheath solution cause to the core liquid to deform into the conical shape and a compound coaxial jet develops at the tip of the cones (Fig. 10).

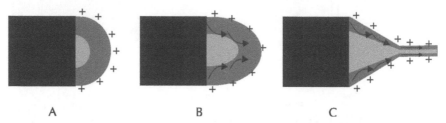

A                                       B                                       C

**FIGURE 10**   Schematic illustration of compound Taylor cone formation (A): surface charges on the sheath solution, (B): viscous drag exerted on the core by the deformed sheath droplet, © Sheath-core compound Taylor cone formed due to continuous viscous drag).

### 9.4.2   NANOFIBERS IN DSSC SOLAR CELL

It can be presumed that the electrospun nanofiber offers high specific surface areas (ranging from hundreds to thousands of m² per gram) and bigger pore sizes than nanoparticle or film. Meanwhile, referring to Section 2.3, the particle-based titanium dioxide layers have low efficiencies due to the high density of grain boundaries, which exist between nanoparticles. The 1D morphology of metal oxide fibers attracts more interest because of the lower density of grain boundaries compared to those of sintered nanoparticles [37] (Table 2).

### 9.4.2.1   ELECTROSPUN SCATTERING LAYER

Ref. [38] compared the effect of $TiO_2$ nanofiber and nanoparticle as scattering layer and indicated the significant enhancement of all photovoltaic specifications. In another attempt ZNO nanofiber used instead of $TiO_2$ nanofiber to form photo anode [39].

**TABLE 2**   The effect of $TiO_2$ nanofiber on cell performance for DSSC solar cell [38–40].

| Solar cell pattern | $V_{oc}$(mV) | $I_{sc}$(mA/cm²) | FF (%) | η (%) | Ref |
|---|---|---|---|---|---|
| FTO\|TiO₂ NP\|N719\|LiI/I₂/ TBT\|Pt\|FTO | 630 | 11.3 | 54 | 3.85 | [38] |
| FTO\|TiO₂ NF\|N719\|LiI/I2/ TBT\|Pt\|FTO | 660 | 14.3 | 53 | 4.9 | [38] |
| FTO\|ZNO NF\|N719\|LiI/I₂/ TBT\|Pt\|FTO | 690 | 2.87 | 44 | 0.88 | [39] |
| FTO\|TiO2 NF: Ag NP\|N719\|LiI/I₂/TBT\|Pt\|FTO | 800 | 7.57 | 55 | 3.3 | [40] |

However, the measured photocurrent density–voltage shows poorer results compared to $TiO_2$. The influence of Ag nanoparticle was also studied showed nearly same fill factor but lower power conversion efficiency compared to neat $TiO_2$ nanofiber scattering layer [40]. Recently, a composite anatase $TiO_2$ nanofibers/nanoparticle electrode was fabricated through electrospinning [41]. This method avoided the mechanical grinding process, and offered a higher surface area, so conversion efficiencies of 8.14% and 10.3% for areas of 0.25 and 0.052 cm², respectively, were reported Hybrid $TiO_2$ nanofibers with moderate multi walled carbon nanotubes (MWCNTs) content also can prolong electron recombination lifetimes [41]. Since MWCNTs can quickly transport charges generated during photocatalysis, the opportunity for charge recombination is reduced. Furthermore, MWCNTs decrease the agglomeration of $TiO_2$ nanoparticles and increase the surface area of $TiO_2$. These advantages make this hybrid electrode a promising candidate for DSSCs.

### 9.4.2.2   NANOFIBER ENCAPSULATED ELECTROLYTE

Nanofiber can be considered as promising candidate for preparation o f solid or semi solid electrolyte. This is mostly because of the inherent long-term instability of electrolyte used in DSSCs usually consists of tri-iodide/iodide redox coupled in organic solvents [42]. Many solid or semisolid viscous electrolytes with low level of penetration to $TiO_2$ layer such as ionic liquids [43], and gel electrolytes [44] used to triumph over these problems. However, nanofiber with may increase the penetration of viscous polymer gel electrolytes through large and controllable pore sizes.

A few research conducted to fabricate the Electrospun PVDF-HFP membrane by electrospinning process from a solution of poly(vinylidenefluoride-cohexafluoropropylene) in a mixture of acetone/N, N-dimethylacetamide to encapsulate electrolyte solution [45, 46]. Although the solar energy-to electricity conversion efficiency of the quasi-solid-state solar cells with the electrospun PVDF-HFP membrane was slightly lower than the value obtained from the conventional liquid electrolyte solar cells, this cell exhibited better long-term durability because of the prevention of electrolyte solution leakage.

### 9.4.2.3   FLEXIBLE NANOFIBER AS COUNTER ELECTRODE

Efficient charge transfer from a counter electrode to an electrolyte is a key process during the operation of dye-sensitized solar cells. One of the greatest flexible counter electrodes could be polyaniline (PAni) nanofibers on graphitized polyimide (GPi) carbon films for use in a tri-iodide reduction. These results are due to the high electrocatalytic activity of the PAni nanofibers and the high conductivity of the flexible GPi film. In combination with a dye-sensitized $TiO_2$ photoelectrode and electrolyte, the photovoltaic device with the PAni counter electrode shows an energy conversion efficiency of 6.85%. Short-term stability tests indicate that the photovoltaic device with the PAni counter electrode approximately preserves its initial performance [47]. The major concern for the application of alternative counter electrodes to conventional platinized TCOs in DSSCs is long-term stability. Many publications indicate that during prolonged exposure in corrosive electrolyte, catalysts will detach from the substrate and deposit onto the surface of the semiconductor photoelectrode (Table 3).

**TABLE 3**   Current–voltage characteristics of the dye synthesized solar cells with various electrodes [47].

| Solar cell pattern | $V_{oc}$(mV) | $I_{sc}$(mA/cm$^2$) | FF (%) | η (%) |
|---|---|---|---|---|
| **FTO\|TiO$_2$ nanoparticle\|N719\|PMII/I2/ TBP \|Pt\|FTO** | 820 | 12.61 | 62.3 | 6.44 |
| **FTO\|TiO$_2$ nanoparticle\|N719\|PMII/I2/ TBP \|PAni\|FTO** | 831 | 12.22 | 62.1 | 6.31 |
| **FTO\|TiO$_2$ nanoparticle\|N719\|PMII/I2/ TBP \|PAni** | 856 | 11.59 | 58.7 | 5.82 |
| **FTO\|TiO$_2$ nanoparticle\|N719\|PMII/I2/ TBP \|PAni\|GPi** | 901 | 9.68 | 28.3 | 2.49 |

## 9.4.2.4   A PROPOSED MODEL FOR DSSC TEXTILE SOLAR CELL USING NANOFIBERS

Choosing proper material and structure for DSSC textile solar cell using previously mentioned nanofiber propose potential candidates for designing of an integrated photovoltaic unit. As can be seen in Fig. 11 a multilayer textile DSSC solar cell composed of a complicated pattern whereas nanofiber is dominant in step-by-step fabrication process.

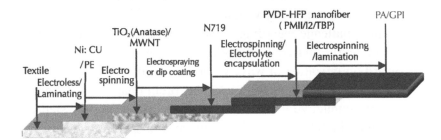

**FIGURE 11**   A proposed model for DSSC textile solar cell using nanofibers in successive layer.

The major concern regarding the DSSC textile solar is $TiO_2$ nanofiber that needs to subject to high temperature for being scattering layer. This strategy is not compatible with other textile element and it is believed that the usage of Anatase $TiO_2$ spinning solution provide the possibility to avoid high temperature treatment. The proposed strategy is under intensive investigation in our laboratory and future results probably mostly illuminate the opportunities and challenges.

Electroless plating is of special interest due to its advantages such as uniform deposition, coherent metal deposition, good electrical and thermal conductivity, shielding effectiveness (SE) and applicability to complex-shaped materials or nonconductors. It can be applied to almost all fiber substances and performed at any step of textile production such as yarn, stock, fabric or cloth [11–13]. Electroless copper plated fabrics are important for manufacturing of conductive textiles due to high conductivity, fast deposition and lower cost [14]. However, the fast oxidation causes this metal to lose its conductivity with time as it is exposed to the air [18]. So, there is need for this material to be coated with a protective material. Nickel could be used as coating material because of the formation of a protective oxide surface in air conditions. Therefore, the process for forming an external protective layer of nickel on the copper surface of PET fabric was done by electroless plating method. The SEM morphology at the top nickel-plated surfaces of metallic fabric are shown in Figs.12(a, b). These indicated that the surfaces were covered with the dense particles and evenness layers which were clearly visible.

The major challenge regarding to the light transmission and outward scattering of textile material after elctroless plating turn the focus of study toward optimization of wave guide profile after textile being conductive using metallic nanoaprticles. The preliminary data of optical properties of conductive textile silk substrate after a step-by-step procedure outlined in Table 4 that shows a semi transparent behavior with 40% reflection and 35% scattering for neat silk fabric. After electroless plating of CU:Ni shows a drastic decreasing of reflectivity. It can mostly attributed to scattering since the transmission insignificantly changes over the optical tests. However, the $Tio_2$ particle and nanofibers assist to increase reflectivity and as it can be expected the transmission share decreases. It means that most of the incident light might be scatters or absorbs in photo anode layers. Experimental observation needs to be substantially carried out to find a transparent conductive layer as a flexible alternative for ITO or FTO glass.

**FIGURE 12** SEM images of the nickel-plated metallic fabric with magnifications 250×(a) and 2500×(b).

**TABLE 4**  The average reflectance and transmittance of silk fabric substrate after layer-by-layer coating with electroless metallic nanoparticle, Titania nanoparticle (NP) and Titania nanofiber.

| Sample | Reflectance, % | Transmission, % |
|---|---|---|
| Silk | 40 | 25 |
| Silk\|CU:Ni | 15 | 24 |
| Silk\|CU:NI\|TiO$_2$NP | 22 | 22 |
| Silk\|CU:NI\|TiO$_2$NP\|TiO2 NF | 28 | 4 |

In spite of all opportunities proposed by DSSC solar cell technology and its bright future in energy conversion sector, but numerous challenges should be addressed for commercialized process. A range of transparent fabric such as Nylon, Polyester, PP and PE have been produced in large scale for various application. Whereas indium tin oxide (ITO) is the prevalent material of choice for conductive transparent substrtate, ITO is less desirable for emerging flexible applications since the brittleness of ITO thin films have a tendency to crack and fail under bending and flexing. Furthermore, ITO deposition requires vacuum coating techniques, and subsequent high-temperature anneal steps that are not compatible with fabric substrates and scalable to high throughput roll-to-roll or printing processes. Although using atomic layer deposition technique in low temperature showed promising potential to overcome the problem, but to best of our knowledge it is extremely expensive process for industrial scale. Therefore, if it would be inevitable to use semi transparent conductive textile layer, developing low band gap excitable dye critically resolve the overall issues regarding to low light transmission intensity.

### 9.4.3  ELECTROSPUN NONWOVEN ORGANIC SOLAR CELL

The idea of generating nonwoven photovoltaic (PV) cloths using organic conducting polymers by electrospinning is quite new and has not been intensively investigated. Based on previously reported PV fiber composed of conjugated hole conducting polymer (*see* Section 9.3.2), a novel methodology was reported to generate a nonwoven organic solar cloth. The fabrication of core–shell nanofibers has been achieved by coelectrospinning of two components such as poly(3-hexyl thiophene) (P3HT) (a conducting polymer) or P3HT/PCBM as the core and poly(vinyl pyrrolidone) (PVP) as the shell using a coaxial electrospinning set up [*see* Section 9.4.1] [48].

Initial measurements of the current density vs. voltage of the P3HT/PCBM solar cloth were carried out and showed current density ($I_{sc}$), open circuit voltage ($V_{oc}$) and fill factor (FF) of the fiber cloth around $3.2 \times 10^{-6}$ mA/cm²–0.12 V and 22.1%, respectively. In addition, a six order of magnitude lower photo conversion efficiency of the fiber cloth around $8.7 \times 10^{-8}$ was observed that might sound disappointing. The low PV parameters of the fiber cloth could be attributed to the following factors:

a)  The fiber cloth processing steps including electrospinning as well as ethanol washing were carried out under ambient conditions.

(b) The thickness of the fiber cloth was ~5 μm compared to the diffusion length of the charge carriers in organic solar cells, which is only several nm.

Therefore, most of the charge carriers were lost in the fiber matrix itself. Drop casted films of the same thickness also showed similar PV parameters.

## 9.4.4   FUTURE PROSPECTS OF ORGANIC NANOFIBER TEXTILE SOLAR

Nanofiber revolutionize the future trend of material selection to enhance the characteristics of textile solar cell. Latest experience regarding to polymer solar cell and low band gap material needs to be considered for ambitious plans with nanofiber morphology. Figure 13 shows a schematic for layer-by-layer hetrojunction textile solar cell according to previously mentioned concerns and promising potential solar absorbing and photoactive material. The usage of anti reflective and protective materials is extremely crucial for industrial scale production. A protective layer will save the organic material from moisture and oxygen. The antireflective layer in solar cells can obstruct the reflection of light and also contribute to the device performance. This is an important point that should be overcome in the case of large-scale production of textile solar cell. The PEDOT:PSS combination with CNTs forms first P-N junction in the form of bi layer nanofiber. The second layer composed of core shell or general bi-layer nanofiber of MDMO/PPV:PCBM, which absorb visible spectrum. The proposed plan can be realized in a large-scale production through a needle-less electrospinning setup. The continuity of process also is not beyond expectation and a range of materials as nanofiber has been already provided on given substrates.

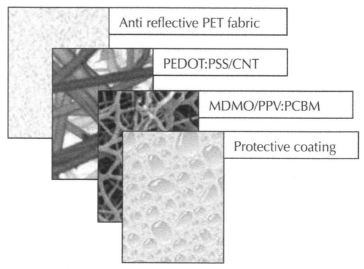

**FIGURE 13**    A general scheme of hetrojunction Organic nanofiber textile solar.

## 9.5  CONCLUDING REMARK

The multilayer solar cell energy conversion unit obeys theories of light scattering, absorption, electron excitation, charge transfer and its compensation. Each layer for specific prescribed functions needs to be intensively investigated for being applicable in diverse circumstances. The commercial solar cell products including silicon either homo or heterojunctions, dye synthesized Ll and organic solar cell subjected to demanding research and reached to high level of maturity. The huge progress in silicon solar cell and other photovoltaic system highlights developing of new generation of flexible solar cell. Current work in first part has a quick glance on variety of solar cells including inorganic, organic and hybrid structures. In overall, regardless of type of solar cell and its corresponding elements, following points have been addressed to show the effect of various parameters on cell performance:

- The wider absorption wavebands in a range of spectrum;
- The thinner the solar cell;
- The lowering the band gap;
- The higher surface area per unit mass;
- The lowering the cathode thicknesses;
- Using antireflective and proactive coating;
- The multi junction cell to cover range of spectrum;
- Using semitransparent wave guide material;
- The lifetime enhancement.

According to these concerns, the recent effort in textile solar cells has been illuminated to find substitute material and routs for integration of nanofibers in textile photovoltaic aspect. The nanofibers showed fascinating performance in DSSC and organic solar cells as scattering, active and also electrolyte layer. The observed performance is attributed to huge surface area and also ability to manipulate the morphology, crystalinity, profile and alignment. In addition making a very thin layers of functional material by a sequential multi step commercial electrospinning process presumably propel the technology to tackle the demanding targets of solar cells including cell efficiency, stability and reliability. Incorporation of DSSC and organic solar cells into textiles is reaching encouraging performances. Stability issues need to be resolved before future commercialization can be envisaged. The mechanical stability of the devices was not limiting the function of the devices prepared. It would seem that low power conversion efficiency much more pertinent than the mechanical stability on the timescale of commercial.

## KEYWORDS

- black dye
- heterojunction solar cells
- polyaniline nanofibers
- silk fabric substrate

## REFERENCES

1. Winterhalter, C. A.; Teverovsky, J.; Wilson; Slade J.; Horowitz, W.; Tierney, E.; Sharma Development of Electronic Textiles to Support Networks, Communications, and Medical Applications in Future, U. S. Military Protective Clothing Systems, *IEEE Trans-actions On Information Technology In Biomedicine,* **2005,** *9(3),* 402–406.
2. Nelson, J. The Physics of Solar Cells. Imperial College Press; **2003.**
3. Miles, R. W.; Forbes, H. I. An overview of state-of-the-art cell development and environmental issues. Progress in Crystal Growth and Characterization of Materials. *Photovoltaic Solar Cells,* **2005,** *51,* 1–42.
4. Günes, S.; Beugebauer, H.; Sariciftci, N S. Conjugated Polymer-based Organic Solar Cells, Chemistry Review **2007,** *107,* 1324–1338.
5. Castafier™. Solar Cells. In: Castafier™ (Ed.), Practical Handbook of Photovoltaics: fundamentals and Applications, Elsevier **2003,** 71–95.
6. Zhu, H.; Wei J.; Wang K.; Wu D. *Applications of carbon materials in photovoltaic solar cells.* Solar Energy Materials and Solar Cells, **2009,** *93(9),* 1461–1470.
7. Shrotriya, V.; Yao, Y.; Moriarty, T, Emery, K.; Yang, Y. Accurate Measurement and Characterization of Organic Solar Cells. *Advanced Functional Material* **2006,** *16,* 2016–2023.
8. Krebs, F. C. Polymer Photovoltaics A Practical Approach: SPIE **2008,** 1–10.
9. Cai, W.; Cao, Y. Polymer solar cells: recent development and possible routes for improvement in the performance. Solar Energy Materials and Solar Cells **2010,** *94,* 114–127.
10. Dutta; Eom S. H.; Lee, S. H.; Synthesis and characterization of triphenylamine flanked thiazole-based small molecules for high performance solution processed organic solar cells. *Organic Electronics* **2012,** *13(2),* 273–282.
11. Soa, S.; Koa, H. M.; Kima, C.; Paeka, S.; Choa, N.; Songb, K.; Leec, JK.; Koa, J. Novel unsymmetrical push–pull squaraine chromophores for solution processed small molecule bulk heterojunction solar cells. *Solar Energy Materials and Solar Cells* **2012,** *98,* 224–232.
12. Lina, Y.; Liua, Y.; Shia, Q.; Hua, W.; Lia, Y.; Zhan, X.; Small molecules based on bithiazole for solution-processed organic solar cells. *Organic Electronics* **2012,** *13(4),* 673–680.

13. Gratzel, M. Photoelectrochemical cells. Nature **2001**, *414*, 338–44.
14. Ferrazza, F. Large size multicrystalline silicon ingots. Proc. E-MRS **2001**, Spring Meeting, Symposium E on Crystalline Silicon Solar Cells. *Solar Energy Material Solar Cells* **2002**, *72*, 77–81.
15. Kisserwan, H. Enhancement of photovoltaic performance of a novel dye, "T18", with ketene thioacetal groups as electron donors for high efficiency dye-sensitized solar cells. Inorganica Chimica Acta **2010**, *363*, 2409–2415.
16. Wei, D. Dye Sensitized Solar Cells. *International J. Molecular Science* **2010**, *11*, 1103–**1113**.
17. Günes, S.; Beugebauer H.; Sariciftci NS. Conjugated Polymer-based Organic Solar Cells, *Chemical Review* **2007**, *107*, 1324–1338.
18. Brabec, C. J.; Dyakonov, V.; Parisi J.; Sariciftci N. S. Organic Photovoltaics Concepts and Realization, 1st ed, Springer, New York, **2003**.
19. Berson, S.; De Bettignies R.; Bailly S.; Guillerez S. Poly(3-hexylthiophene) Fibers for Photovoltaic Applications. Advanced Funcional Material **2007**, *17*, 1377–1384.
20. Gonzalez, R.; Pinto, N. J. Electrospun poly(3-hexylthiophene-2, 5-diyl) Fiber Field Effect Trans-istor. *Synthetic Metals* **2005**, *151*, 275–278.
21. Bahners, T.; Schlosser, U.; Gutmann, R.; Schollmeyer, E. Textile solar light collectors based on models for polar bear hair. *Solar Energy Materials and Solar Cells* **2008**, *92*, 1661–1667.
22. Tributsch, H.; Goslowski H.; Ku U.; Wetzel H. Light collection and solar sensing through the polar bear pelt, *Solar Energy Material* **1990**, *21*, 219–236.
23. Liu, J.; Namboothiry M. A. G.; Carroll DL. Fiber based Architectures for Organic Photovoltaics. *Applied Physics Letter* **2007**, *90*, 063501.
24. O'Connor, B.; Pipe K; Shtein M. Fiber Based Organic Photovoltaic Devices. *Applied Physics Letter* **2008**, *92*, 193306.
25. Bedeloglu, A.; Demir A.; Bozkurt Y.; Sariciftci N. S. A Photovoltaic Fiber Design for Smart Textiles. *Textile Research Journal* **2007**, *80(11)*, 1065–1074.
26. Ouyang, J.; Chu C. W.; Chen F. C.; Xu Q.; Yang Y. High-conductivity Poly (3, 4-ethylenedioxythiophene*)*, poly(stirene sulfonate) Film and its Application in Polymer Optoelectronic Devices, *Advanced Functional Materials* **2005**, *15*, 203–208.
27. Kushto, G. P.; Kim W.; Kafafi, Z. H.; Flexible Organic Photovoltaics using Conducting Polymer Electrodes, *Applied Physics Letters* **2005**, *86*, 093502.
28. Huang, J.; Wang X.; Kim Y.; deMello, A. J.; Bradley, D.D. C.; De Mello, J. C.; High Efficiency Flexible ITO-free Polymer/fullerene Photodiodes, *Physical Chemistry Chemical Physics* **2006**, *8*, 3904–3908.
29. Tvingstedt, K.; Inganäs, O. Electrode Grids for ITO-free Organic Photovoltaic Devices, Advanced. Materials **2007**, *19*, 2893–2897.
30. Perzon, E.; Wang, X.; Admassie, S.; Inganäs O.; Andersson M. R.; An Alternating Low Band-gap Polyfluorene for Optoelectronic Devices. *Polymer* **2006**, *47*, 4261–4268
31. Campos, L. M.; Tontcheva, A.; Günes, S.; Sonmez, G.; Neugebauer, H.; Sariciftci, N. S.; Wudl, F. Extended Photocurrent Spectrum of a Low Band Gap Polymer in a Bulk eterojunction Solar Cell. *Chemistry of Materials* **2005**, *17*, 4031–4033.
32. Scharber, M.S.; Mühlbacher, D.; Koppe, M.; Denk; Waldauf, C.; Heeger, A. J.; Brabec, C. J. *Design Rules for Donors in Bulk-heterojunction Solar Cells Towards 10 % Energy-conversion Efficiency*, Advanced Materials **2006**, *18*, 789–794.

33. Krebs, F.C.; Biancardo, M.; Jensen B.W.; Spanggard, H.; Alstrup, J. Strategies for incorporation of polymer photovoltaics into garments and textiles. Solar Energy Materials and Solar Cells **2006**, *90,* 1058–1067.

34. Li, D.; Wang, Y.; Xia, Y. Electrospinning Nanofibers as Uniaxially Aligned Arrays and Layer-by-layer Stacked Films. Advanced Materials **2004**, *16,* 14, 361–366.

35. Yu, J.H.; Fridrikh, S.V.; Rutledge, G.C. Production of Submicrometer Diameter Fibers by Two-fluid Electrospinning. Advanced Materials **2004**, *16, 17,* 1562–1566.

36. Li D, Xia Y.; Direct Fabrication of Composite and Ceramic Hollow Nanofibers by Electrospinning, Nano Letters **2004**, 4, 5, 933–938.

37. Chuangchote S.; Sagawa T.; Yoshikawa S.; Applied Physics Letters **2008**, 93, 033310.

38. Zhao X.; Lin H.; Li X.; Li J. The application of freestanding titanate nanofiber paper for scattering layers in dye-sensitized solar cells. Materials Letters **2011**, *65,* 1157–1160.

39. Li S.; Zhang X, Jiao X.; Lin H.; One-step large-scale synthesis of porous ZnO nanofibers and their application in dye-sensitized solar cells. Materials Letters **2011**, 65, 2975–2978.

40. Li J.; Chen X, Ai N.; Hao J.; Chen Q.; Strauf S.; Shi Y, Silver nanoparticle doped TiO2 nanofiber dye sensitized solar cells. Chemical Physics Letters **2011**, *514,* 141–145.

41. Hu GJ.; Meng XF.; Feng XY.; Ding YF.; Zhang SM.; Yang MS.; Anatase TiO$_2$ nanoparticles/carbon nanotubes nanofibers: preparation, characterization and photocatalytic properties. J. Material Science **2007**, *42,* 7162–7170.

42. Kubo W.; Kitamura T.; Hanabusa K.; Wada Y.; Yanagida S. Quasi-solid-state dye-sensitized solar cells using room temperature molten salts and a low molecular weight gelator. Chemical Communication **2002**, 374–375.

43. Wang; Zakeeruddin SM.; Comte; Exnar I.; Gratzel M.; Gelation of Ionic Liquid-based Electrolytes with Silica Nanoparticles for Quasi-solid-state Dye-sensitized Solar Cells, J. American Chemical society **2003**, *125,* 1166–1167.

44. Wang,; Zakeeruddin, S. M.; Moser, J. E.; Nazeeruddin, M. E.; Sekiguchi, T.; Gratzel, M.; A stable quasi-solid-state dye-sensitized solar cell with an amphiphilic ruthenium sensitizer and polymer gel electrolyte. Natural Material **2003**, 2, 402–407.

45. Park, S. H.; Kim, J. U.; Lee, S. Y.; Lee, W. K.; Lee, J. K.; Kim, M. R. Dye-sensitized Solar Cells Using Polymer Electrolytes Based on Poly(vinylidene fluoride-hexafluoro propylene) Nanofibers by Electrospinning Method. J. Nanoscience and Nanotechnology 2008, 8, 4889–4894.

46. Kim J. U.; Park, S. H.; Choi, H. J.; Lee, W. K.; Lee, J. K.; Kim, M. ;R. Effect of electrolyte in electrospun poly(vinylidene fluoride-*co*hexafluoropropylene) nanofibers on dye-sensitized solar cells Solar Energy Materials and Solar Cells **2009**, 93–803–807.

47. Chen, J.; Lia, B.; Zheng, J.; Zhao, J.; Zhu, Z.; Polyaniline nanofiber/carbon film as flexible counter electrodes in platinum-free dye-sensitized solar cells. Electrochimica Acta 2011, 56, 4624–4630.

48. Sundarrajan, S.; Murugan, R.; Nair, A.S.; Ramakrishna, S.; Fabrication of P3HT/PCBM solar cloth by electrospinning technique. Materials Letters **2010**, *64,* 2369–2372.

# CHAPTER 10

# POTENTIAL APPLICATIONS OF METAL-ORGANIC FRAMEWORKS IN TEXTILES

M. HASANZADEH and B. HADAVI MOGHADAM

## CONTENTS

## 10.1   INTRODUCTION

Recently, the application of nanostructured materials has garnered attention, due to their interesting chemical and physical properties. Application
of nanostructured materials on the solid substrate such as fibers brings
new properties to the final textile product [1]. Metal-organic frameworks
(MOFs) are one of the most recognized nanoporous materials, which can
be widely used for modification of fibers. These relatively crystalline materials consist of metal ions or clusters (named secondary building units,
SBUs) interconnected by organic molecules called linkers, which can possess one, two or three dimensional structures [2–10]. They have received
a great deal of attention, and the increase in the number of publications
related to MOFs in the past decade is remarkable (Fig. 1).

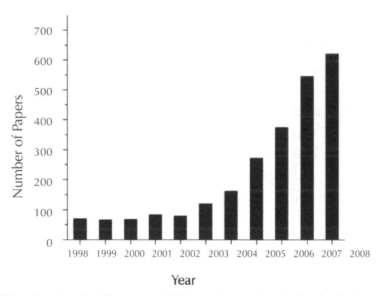

**FIGURE 1**   Number of publications on MOFs over the past decade, showing the increasing
research interest in this topic.

These materials possess a wide array of potential applications in many
scientific and industrial fields, including gas storage [11, 12], molecular
separation [13], catalysis [14], drug delivery [15], sensing [16], and others.
This is due to the unique combination of high porosity, very large surface

areas, accessible pore volume, wide range of pore sizes and topologies, chemical stability, and infinite number of possible structures [17, 18].

Although other well-known solid materials such as zeolites and active carbon also show large surface area and nanoporosity, MOFs have some new and distinct advantages. The most basic difference of MOFs and their inorganic counterparts (e.g., zeolites) is the chemical composition and absence of an inaccessible volume (called dead volume) in MOFs [10]. This feature offers the highest value of surface area and porosities in MOFs materials [19]. Another difference between MOFs and other well-known nanoporous materials such as zeolites and carbon nanotubes is the ability to tune the structure and functionality of MOFs directly during synthesis [17].

The first report of MOFs dates back to 1990, when Robson introduced a design concept to the construction of 3D MOFs using appropriate molecular building blocks and metal ions. Following the seminal work, several experiments were developed in this field such as work from Yaghi and O'Keeffe [20].

In this review, synthesis and structural properties of MOFs are summarized and some of the key advances that have been made in the application of these nanoporous materials in textile fibers are highlighted.

## 10.2   SYNTHESIS OF MOFS

MOFs are typically synthesized under mild temperature (up to 200°C) by combination of organic linkers and metal ions (Fig. 2) in solvothermal reaction [2, 21].

Recent studies have shown that the character of the MOF depends on many parameters including characteristics of the ligand (bond angles, ligand length, bulkiness, chirality, etc.), solubility of the reactants in the solvent, concentration of organic link and metal salt, solvent polarity, the pH of solution, ionic strength of the medium, temperature and pressure [2, 21].

In addition to this synthesis method, several different methodologies are described in the literature such as ball-milling technique, microwave irradiation, and ultrasonic approach [22].

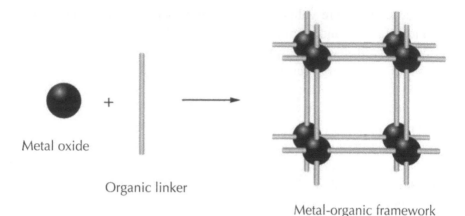

Metal oxide

Organic linker

Metal-organic framework

**FIGURE 2**    Formation of metal organic frameworks.

Post-synthetic modification (PSM) of MOFs opens up further chemical reactions to decorate the frameworks with molecules or functional groups that might not be achieved by conventional synthesis. In situations that presence of a certain functional group on a ligand prevents the formation of the targeted MOF, it is necessary to first form a MOF with the desired topology, and then add the functional group to the framework [2].

## 10.3    STRUCTURE AND PROPERTIES OF MOFS

When considering the structure of MOFs, it is useful to recognize the secondary building units (SBUs), for understanding and predicting topologies of structures [3]. Figure 3 shows the examples of some SBUs that are commonly occurring in metal carboxylate MOFs. Figure 3(a–c) illustrates inorganic SBUs include the square paddlewheel, the octahedral basic zinc acetate cluster, and the trigonal prismatic oxo-centered trimer, respectively. These SBUs are usually reticulated into MOFs by linking the carboxylate carbons with organic units [3]. Examples of organic SBUs are also shown in Fig. 3(d–f).

(a)                    (b)                    (C)

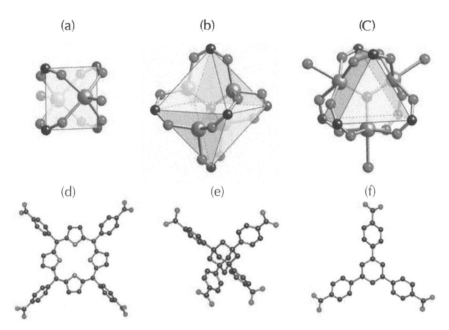

(d)                    (e)                    (f)

**FIGURE 3**   Structural representations of some SBUs, including (a–c) inorganic, and (b–f) organic SBUs. (Metals are shown as blue spheres, carbon as black spheres, oxygen as red spheres, nitrogen as green spheres).

It should be noted that the geometry of the SBU is dependent on not only the structure of the ligand and type of metal used, but also the metal to ligand ratio, the solvent, and the source of anions to balance the charge of the metal ion [2].

A large number of MOFs have been synthesized and reported by researchers to date. Isoreticular metal-organic frameworks (IRMOFs) denoted as IRMOF-$n$ ($n$ = 1 through 7, 8, 10, 12, 14, and 16) are one of the most widely studied MOFs in the literature. These compounds possess cubic framework structures in which each member shares the same cubic topology [3, 21]. Figure 4 shows the structure of IRMOF-1 (MOF-5) as simplest member of IRMOF series.

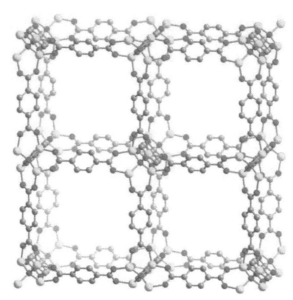

**FIGURE 4** Structural representation of IRMOF-1. (Yellow, gray, and red spheres represent Zn, C, and O atoms, respectively).

## 10.4   APPLICATION OF MOFS IN TEXTILES

### 10.4.1   INTRODUCTION

There are many methods of surface modification, among which nanostructure based modifications have created a new approach for many applications in recent years. Although MOFs are one of the most promising nanostructured materials for modification of textile fibers, only a few examples have been reported to data. In this section, the first part focuses on application of MOFs in nanofibers and the second part is concerned with modifications of ordinary textile fiber with these nanoporous materials.

### 10.4.2   NANOFIBERS

Nanofibrous materials can be made by using the electrospinning process. Electrospinning process involves three main components including syringe filled with a polymer solution, a high voltage supplier to provide the required electric force for stretching the liquid jet, and a grounded collection plate to hold the nanofiber mat. The charged polymer solution forms a liquid jet that is drawn towards a grounded collection plate. During the jet movement to the collector, the solvent evaporates and dry fibers deposited as randomly oriented structure on the surface of a collector [23–28]. The schematic illustration of conventional electrospinning setup is shown in Fig. 5.

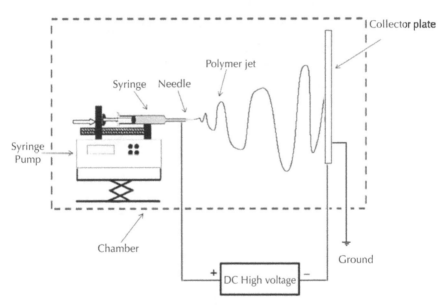

**FIGURE 5**   Schematic illustration of electrospinning set up.

At the present time, synthesis and fabrication of functional nanofibers represent one of the most interesting fields of nanoresearch. Combining the advanced structural features of metal-organic frameworks with the fabrication technique may generate new functionalized nanofibers for more multiple purposes.

However, there has been great interest in the preparation of nanofibers, the studies on metal-organic polymers are rare. In the most recent investigation in this field, the growth of MOF (MIL-47) on electrospun polyacrylonitrile (PAN) mat was studied using in situ microwave irradiation [18]. MIL-

47 consists of vanadium cations associated to six oxygen atoms, forming chains connected by terephthalate linkers (Fig. 6).

It should be mentioned that the conversion of nitrile to carboxylic acid groups is necessary for the MOF growth on the PAN nanofibers surface.

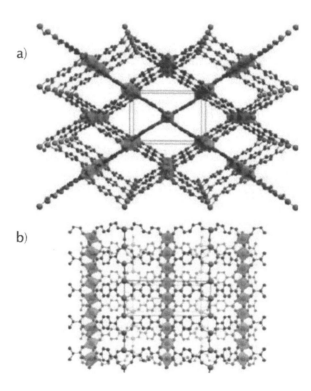

a)

b)

**FIGURE 6** MIL-47 metal-organic framework structure: view along the b axis (a) and along the c axis (b).

The crystal morphology of MIL-47 grown on the electrospun fibers illustrated that after only 5 s, the polymer surface was partially covered with small agglomerates of MOF particles. With increasing irradiation time, the agglomerates grew as elongated anisotropic structures (Fig. 7) [18].

**FIGURE 7**   SEM micrograph of MIL-47 coated PAN substrate prepared from electrospun nanofibers as a function of irradiation time: (a) 5 s, (b) 30 s, (c) 3 min, and (d) 6 min.

It is known that the synthesis of desirable metal-organic polymers is one of the most important factors for the success of the fabrication of metal-organic nanofibers [29]. Among several novel microporous metal-organic polymers, only a few of them have been fabricated into metal-organic fibers.

For example, new acentric metal-organic framework was synthesized and fabricated into nanofibers using electrospinning process [29]. The two dimensional network structure of synthesized MOF is shown in Fig. 8. For this purpose, MOF was dissolved in water or DMF and saturated MOF solution was used for electrospinning. They studied the diameter and morphology of the nanofibers using an optical microscope and a scanning electron microscope (Fig. 9). This fiber display diameters range from 60 nm to 4 μm.

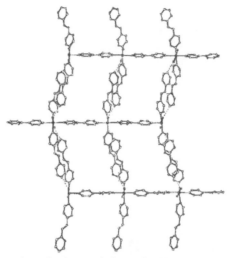

**FIGURE 8**    Representation of polymer chains and network structure of MOF.

**FIGURE 9**    SEM micrograph of electrospun nanofiber.

In 2011, Kaskel et al. [30], reported the use of electrospinning process for the immobilization of MOF particles in fibers. They used HKUST-1 and MIL-100(Fe) as MOF particles, which are stable during the electro-spinning process from a suspension. Electrospun polymer fibers contain-ing up to 80 wt% MOF particles were achieved and exhibit a total acces-sible inner surface area. It was found that HKUST-1/PAN gives a spider web-like network of the fibers with MOF particles like trapped flies in it, however, HKUST-1/PS results in a pearl necklace-like alignment of the crystallites on the fibers with relatively low loadings.

## 10.4.3   ORDINARY TEXTILE FIBERS

Some examples of modification of fibers with metal-organic frameworks have verified successful. For instance, in the study on the growth of $Cu_3(BTC)_2$ (also known as HKUST-1, BTC=1,3,5-benzenetricarboxylate) MOF nanostructure on silk fiber under ultrasound irradiation, it was demonstrated that the silk fibers containing $Cu_3(BTC)_2$ MOF exhibited high antibacterial activity against the gram-negative bacterial strain *E. coli* and the gram-positive strain *S. aureus* [1]. The structure and SEM micrograph of $Cu_3(BTC)_2$ MOF is shown in Fig. 10.

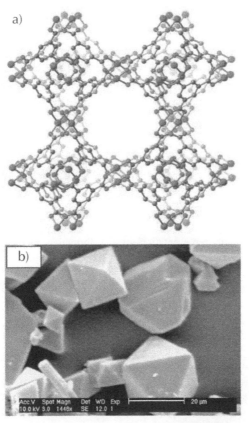

**FIGURE 10**   (a) The unit cell structure and (b) SEM micrograph of the $Cu_3(BTC)_2$ metal-organic framework. (Green, gray, and red spheres represent Cu, C, and O atoms, respectively).

$Cu_3(BTC)_2$ MOF has a large pore volume, between 62% and 72% of the total volume, and a cubic structure consists of three mutually perpendicular channels [32].

The formation mechanism of $Cu_3(BTC)_2$ nanoparticles upon silk fiber is illustrated in Fig. 11. It is found that formation of $Cu_3(BTC)_2$ MOF on silk fiber surface was increased in presence of ultrasound irradiation. In addition, increasing the concentration cause an increase in antimicrobial activity [1]. Figure 12 shows the SEM micrograph of $Cu_3(BTC)_2$ MOF on silk surface.

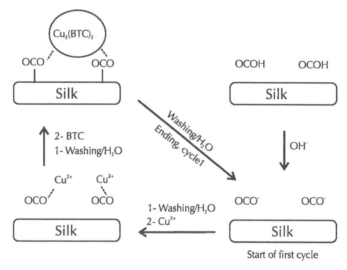

**FIGURE 11** Schematic representation of the formation mechanism of $Cu_3(BTC)_2$ nanoparticles upon silk fiber.

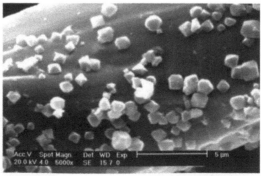

**FIGURE 12** SEM micrograph of $Cu_3(BTC)_2$ crystals on silk fibers.

The FT-IR spectra of the pure silk yarn and silk yarn containing MOF (CuBTC-silk) are shown in Fig. 13. Owing to the reduction of the C=O bond, which is caused by the coordination of oxygen to the $Cu^{2+}$ metal center (Fig. 11), the stretching frequency of the C=O bond was shifted to lower wavenumbers (1,654 $cm^{-1}$) in comparison with the free silk (1,664 $cm^{-1}$) after chelation [1].

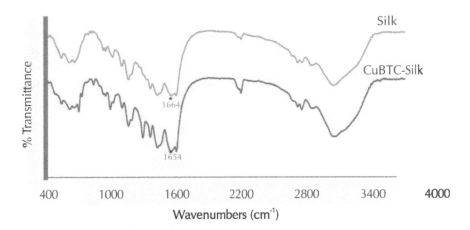

**FIGURE 13**    FT-IR spectra of the pure silk yarn and silk yarn containing $Cu_3(BTC)_2$.

In another study, $Cu_3(BTC)_2$ was synthesized in the presence of pulp fibers of different qualities [33]. The following pulp samples were used: a bleached and an unbleached kraft pulp, and chemithermomechanical pulp (CTMP).

All three samples differed in their residual lignin content. Indeed, owing to the different chemical composition of samples, different results regarding the degree of coverage were expected. The content of $Cu_3(BTC)_2$ in pulp samples, $k$-number, and single point BET surface area are shown in Table 1. $k$-number of pulp samples, which is indicates the lignin content indirectly, was determined by consumption of a sulfuric permanganate solution of the selected pulp sample [33].

**TABLE 1**   Some characteristics of the pulp samples.

| Pulp sample | MOF content[a] (wt.%) | k-number[b] | Surface area[c] (m² g⁻¹) |
|---|---|---|---|
| CTMP | 19.95 | 114.5 | 314 |
| Unbleached kraft pulp | 10.69 | 27.6 | 165 |
| Bleached kraft pulp | 0 | 0.3 | 10 |

[a]Determined by thermogravimetric analysis.
[b]Determined according to ISO 302.
[c]Single point BET surface area calculated at $p/p_0$=0.3 bar.

It is found that CTMP fibers showed the highest lignin residue and largest BET surface area. As shown in the SEM micrograph (Fig. 14), the crystals are regularly distributed on the fiber surface. The unbleached kraft pulp sample provides a slightly lower content of MOF crystals and BET surface area with 165 $m_2\,g^{-1}$. Moreover, no crystals adhered to the bleached kraft pulp, which was almost free of any lignin.

**FIGURE 14**   SEM micrograph of $Cu_3(BTC)_2$ crystals on the CTMP fibers.

## 10.5   CONCLUSION

New review on feasibility and application of several kinds of metal-organic frameworks on different substrate including nanofiber and ordinary fiber was investigated. Based on the researcher's results, the following conclusions can be drawn:

1. MOF, as new class of nanoporous materials, can be used for modification of textile fibers.
2. These nanostructured materials have many exciting characteristics such as large pore sizes, high porosity, high surface areas, and wide range of pore sizes and topologies.
3. Although tremendous progress has been made in the potential applications of MOFs during past decade, only a few investigations have reported in textile engineering fields.
4. Morphological properties of the MOF/fiber composites were defined; the most advantageous, particle size distribution was shown.
5. It is concluded that the MOFs/fiber composite would be good candidates for many technological applications, such as gas separation, hydrogen storage, sensor, and others.

## KEYWORDS

- metal-organic frameworks
- nanofibers
- porous materials
- textiles

## REFERENCES

1. Abbasi, A. R.; Akhbari K.; Morsali A. Dense coating of surface mounted CuBTC metal-organic framework nanostructures on silk fibers, prepared by layer-by-layer method under ultrasound irradiation with antibacterial activity. *Ultrasonics Sonochemistry*, **2012**, *19*, 846–852.
2. Kuppler, R. J.; Timmons, D. J.; Fang Q.-R.; Li J.-R.; Makal, T. A.; Young, M. D.; Yuan D.; Zhao D.; Zhuang W.; Zhou H.-C. Potential applications of metal-organic frameworks. *Coordination Chemistry Reviews*, **2009**, *253*, 3042–3066.
3. Rowsell, J. L. C.; Yaghi, O. M.: Metal-organic frameworks: a new class of porous materials. *Microporous and Mesoporous Materials*, **2004**, *73*, 3–14.

4. An J.; Farha, O. K.; Hupp, J. T.; Pohl E.; Yeh J. I.; Rosi, N. L. Metal-adeninate vertices for the construction of an exceptionally porous metal-organic framework. *Nature communications*, DOI: 10.1038/ncomms1618, **2012.**

5. Morris W.; Taylor, R. E.; Dybowski C.; Yaghi, O. M.; Garcia-Garibay, M. A. Framework mobility in the metal-organic framework crystal IRMOF-3: evidence for aromatic ring and amine rotation. *J. Molecular Structure*, **2011,** 1004, 94–101.

6. Kepert, C. J. Metal-organic framework materials. In 'Porous Materials' (eds.: by Bruce, D. W.; O'Hare D. and Walton, R. I.) John Wiley and Sons, Chichester, **2011.**

7. Rowsell, J. L. C.; Yaghi, O. M. Effects of functionalization, catenation, and variation of the metal oxide and organic linking units on the low-pressure hydrogen adsorption properties of metal-organic frameworks. *J. the American Chemical Society*, **2006,** *128,* 1304–1315.

8. Rowsell, J. L. C.; Yaghi, O. M. Strategies for hydrogen storage in metal-organic frameworks. *Angewandte Chemie International Edition*, **2005,** *44,* 4670–4679.

9. Farha, O. K.; Mulfort, K. L.; Thorsness, A. M.; Hupp, J. T. Separating solids: purification of metal-organic framework materials. *J. American Chemical Society*, **2008,** *130,* 8598–8599.

10. Khoshaman, A. H. Application of electrospun thin films for supra-molecule based gas sensing. M.Sc. thesis, Simon Fraser University. **2011.**

11. Murray, L. J.; Dinca M.; Long, J. R. Hydrogen storage in metal-organic frameworks. *Chemical Society Reviews*, **2009,** *38,* 1294–1314.

12. Collins, D. J.; Zhou H.-C. Hydrogen storage in metal-organic frameworks. *J. Materials Chemistry*, **2007,** *17,* 3154–3160.

13. Chen, B.; Liang C.; Yang J.; Contreras, D. S.; Clancy, Y. L.; Lobkovsky, E. B.; Yaghi, O. M.; Dai S. A microporous metal-organic framework for gas-chromatographic separation of alkanes. *Angewandte Chemie International Edition*, **2006,** *45,* 1390–1393.

14. Lee, J. Y.; Farha, O. K.; Roberts J.; Scheidt, K. A.; Nguyen, S. T.; Hupp, J. T. Metal-organic framework materials as catalysts. *Chemical Society Reviews*, **2009,** *38,* 1450–1459.

15. Huxford, R. C.; Rocca, J. D.; Lin W. Metal-organic frameworks as potential drug carriers. *Current Opinion in Chemical Biology*, **2010,** *14,* 262–268.

16. Suh M.; Cheon, Y. E.; Lee E. Y. Syntheses and functions of porous metallosupramolecular networks. *Coordination Chemistry Reviews*, **2008,** *252,* 1007–1026.

17. Keskin S.; Kızılel S.: Biomedical applications of metal organic frameworks. *Industrial and Engineering Chemistry Research.* **2011,** *50,* 1799–1812.

18. Centrone A.; Yang Y.; Speakman S.; Bromberg L.; Rutledge, G. C.; Hatton, T. A. Growth of metal-organic frameworks on polymer surfaces. *J. American Chemical Society*, **2010,** *132,* 15687–15691.

19. Wong-Foy, A. G.; Matzger, A. J.; Yaghi, O. M. Exceptional $H_2$ saturation uptake in microporous metal-organic frameworks. *J. the American Chemical Society*, **2006,** *128,* 3494–3495.

20. Farrusseng, D. Metal-organic frameworks: a. Wiley-VCH, Weinheim **2011.**

21. Rosi, N. L.; Eddaoudi M.; Kim J.; O'Keeffe M.; Yaghi, O. M. Advances in the chemistry of metal-organic frameworks. *CrystEngComm*, **2002,** *4,* 401–404.

22. Zou R.; Abdel-Fattah, A. I.; Xu H.; Zhao Y.; Hickmott, D. D.: *Cryst Eng Comm,* **2010,** *12,* 1337–1353.

23. Reneker, D. H.; Chun I. Nanometer diameter fibers of polymer, produced by electrospinning, *Nanotechnology*, **1996**, *7*, 216–223.
24. Shin, Y. M.; Hohman, M. M.; Brenner M.; Rutledge, G. C. Experimental characterization of electrospinning: the electrically forced jet and instabilities. *Polymer*, **2001**, *42*, 9955–9967
25. Reneker, D. H.; Yarin, A. L.; Fong H.; Koombhongse S. Bending instability of electrically charged liquid jets of polymer solutions in electrospinning, *J. Applied Physics*, **2000**, *87*, 4531–4547.
26. Zhang S.; Shim, W. S.; Kim J. Design of ultra-fine nonwovens via electrospinning of Nylon 6: spinning parameters and filtration efficiency, *Materials and Design*, **2009**, *30*, 3659–3666.
27. Yördem, O. S.; Papila M.; Menceloğlu, Y. Z. Effects of electrospinning parameters on polyacrylonitrile nanofiber diameter: an investigation by response surface methodology. *Materials and Design*, **2008**, *29*, 34–44.
28. Chronakis, I. S. Novel nanocomposites and nanoceramics based on polymer nanofibers using electrospinning process-A review. *J. Materials Processing Technology*, **2005**, *167*, 283–293
29. Lu, J. Y.; Runnels, K. A.; Norman, C.: A new metal-organic polymer with large grid acentric structure created by unbalanced inclusion species and its electrospun nanofibers. *Inorganic Chemistry*, **2001**, *40*, 4516–4517.
30. Rose, M.; Böhringer, B.; Jolly, M.; Fischer, R.; Kaskel, S. MOF processing by electrospinning for functional textiles. *Advanced Engineering Materials*, **2011**, 13, 356–360.
31. Basu, S.; Maes, M.; Cano-Odena, A.; Alaerts, L.; De Vos, D. E.; Vankelecom, I. F. J.: solvent resistant nanofiltration (SRNF) membranes based on metal-organic frameworks. *J. Membrane Science*, **2009**, *344*, 190–198.
32. Hopkins, J. B.: infrared spectroscopy of $H_2$ trapped in metal organic frameworks. Thesis, B. A.; Oberlin College Honors **2009.**
33. Küsgens, Siegle S.; Kaskel S. Crystal growth of the metal-organic framework $Cu_3(BTC)_2$ on the surface of pulp fibers. *Advanced Engineering Materials*, **2009**, *11*, 93–95.

# CHAPTER 11

# SPECIAL TOPICS

# SECTION I: BIOMASS LOGISTICS: THE KEY CHALLENGE OF MINIMIZING SUPPLY COSTS

INA EHRHARDT, HOLGER SEIDEL, CHRISTIAN BLOBNER, and E. H. MICHAEL SCHENK

## CONTENTS

### SECTION I

### SECTION II

## 11.1  INTRODUCTION

More than 53% of the renewable energy consumed in Germany is generated from biomass, predominantly wood, and the trend is upward. Research has focused on the utilization of woody biomass and neglected the issues and challenges confronting the actors in the biomass value added and supply chain. Biomass suppliers and consumers are usually unaware of the share of total supply costs incurred by complex logistics and indirect costs. Hence, holistic logistics concepts and bases for planning are needed to supply biomass cost effectively and sustainably as cross-company value added.

Supported by such instruments as the German Renewable Energy Act (EEG) and other European and national support programs, the utilization of renewable energies is increasingly growing in importance. Biomass already supplies around 70% of the energy obtained from renewable energy sources [1]. Whereas biomass (solid biofuels, liquid and gaseous fuels and biowaste) only accounts for 19.8% and thus a small part of the renewable power produced, it is the source of 91.5% of the renewable heat produced [2]. The raw material wood constitutes the most significant proportion of solid biofuels used to supply heat.

Woody biomass generally denotes thinnings, forest wood residues and wood from landscape conservation (roadside vegetation, natural copses, prunings, driftwood). However, energy wood from plantations, industrial and other wood waste wood are also woody biomass.

The utilization of woody biomass is generating competition for this resource now grown scarce. Utilizing woody waste and residues from agriculture, forestry and landscape conservation represents one solution to this conflict. In the Best4VarioUse project, the Fraunhofer IFF is presently researching issues surrounding the supply and utilization of wood biomass residues. Together with 15 partners from Germany and the Spanish region of Valencia, the Fraunhofer IFF is researching the limitations on organizing economically and ecologically efficient biomass supply chains and the implications for concepts for regional utilization.

Biogenic wastes and residues from biotope and landscape conservation, agriculture and forestry contain characteristic amounts of woody biomass dependent on their provenance and composition. (*see* Table 1 for a summary of forestry, agriculture and landscape conservation actions that produce woody biomass waste and residues.) So far, these residues have remained largely unused. Efficiently using raw materials hitherto classified as waste in energy value chains will necessitate implementing and, where necessary, modifying combined methods and technologies so that the material flows are economically and ecologically efficient. The low local biomass yields and wide dispersion of consumers render the great complexity of the logistics of the biomass supply a key challenge. Efficient planning and organization of the logistical structures and exact knowledge of the direct and indirect costs incurred in the supply chain are the basis for the future development of the utilization of biomass residues and wastes. Residues and wastes will only become a true alternative once they can be supplied at reasonable prices and in proper qualities. Moreover, the cost drivers in the supply chain will have to be identified in order to cuts costs.

**TABLE 1**    Actions producing woody biomass waste and residues.

| Domain | Action | | | | |
|---|---|---|---|---|---|
| **Forest** | **Clearance after damage** | **Regular thinning/ selective felling [3]** | **Micro-development** | **Firebreaks** | **Cultivation of succession lands [4]** |
| Landscape conservation | Tree care | | Preservation of open landscapes | Roadside vegetation | |
| Agriculture | Hedgerow care | Deshrubbing | Pruning/ tree care | Cultivation of abandoned land [5] | Clearance |
| Other | Construction site clearance | | | Traffic safety [6] | |

## 11.2   FACTORS INFLUENCING THE BIOMASS SUPPLY CHAIN

Methods and tools of supply chain management (SCM) provide the basic principles and means to analyze logistical issues related to the value added chain and the organization of the biomass supply chain. These tools raise companies' awareness of the overall operation and thus increase cross-company value added. SCM revolves around cross-company coordination of material and information flows throughout the entire value added process, from raw material production to individual stages of processing up through the end consumer, with the goal of organizing the overall operation so that both time and cost have been optimized. Logistics interfaces are crucial to a supply chain's smooth operation. Hence the goal must be to optimize the interfaces by eliminating unnecessary activities and unnecessary resource consumption at system interfaces in order to optimize costs.

A systematic analysis is needed to optimize biomass value added ecologically and economically. First, the operational stages of a typical supply chain are analyzed. Then, select factors that influence auf direct and indirect supply and logistics costs are mapped.

Biomass logistics operations subsume every stage from harvesting to final delivery to processing. Since upstream and preparatory forest stand planning and tending operations or operations in a wood processing plant are not considered direct elements, they are only analyzed tangentially. The planning, control, execution, monitoring and billing of all biomass harvesting, storage, processing and transport operations and even related information flows are also part of biomass logistics. The typical overall supply chain from the "forest to the factory" encompasses the individual operations of harvesting, extraction, assortment formation, storage and transport (*see* Fig. 1).

**FIGURE 1**    Biomass supply chain.

Numerous factors, which can be categorized as external or internal influences, determine the amount of the costs incurred by supplying biomass. They affect indirect or direct costs related to operations or raw material (*see* ). For instance, hauling distances and transfer points (forest roads and temporary storage facilities) affect transport costs. The number of competing providers in a region or the prices of competing fossil fuels also affect the price of biomass, which must not only cover supply costs but also furnish the actors involved in the operation a profit margin [7].

**TABLE 2**   Morphology of factors influencing costs incurred.

| Factors | |
| --- | --- |
| Influencing direct costs | Influencing indirect costs |
| Internal | External |
| Operations-related | Raw material-related |

## 11.3   BIOMASS SUPPLY COSTS

Generally, biomass supply or logistics costs are incurred by the operations to supply the raw material from its site of production to consumers. Five different types of logistics costs have been distinguished [8].

1. System and control costs, i.e., costs incurred by the organization, planning and monitoring of the material flow and by production program planning, material planning, order processing and production control.
2. Inventory costs, i.e., costs incurred by maintaining and financing stocks.
3. Storage costs, i.e., costs incurred by maintaining storage capacities and storage and retrieval operations.
4. Transport costs, i.e., costs incurred by internal and external plant transport.
5. Handling costs, i.e., costs incurred by packaging and picking partial shipments from a total ordered shipment.

Supply chain costs must be identified for analysis and optimization. Part of the costs, e.g., harvesting costs, can be measured directly, thus eliminating the need for detailed research since they are relatively obvious. Indirect costs, for example, administrative or communication costs, are more difficult to identify. A study conducted by the Fraunhofer IFF in Magdeburg revealed that indirect supply costs are usually not transparent, thus neither quantifiable nor qualifiable and Therefore, frequently not entered into costing and pricing. Consequently, biomass providers are aware in the planning phase of the direct (procurement) costs of services such as harvesting, skidding or transport but not the total costs that offset the "value" of the particular biomass supplied. Thus, marketing and assortment decisions are usually based only on the direct supply costs and the obtainable market price and factor in no other criteria and variables. Especially in Germany, the large number of actors and differing contracts contained in the value added chain significantly affect the amount of the indirect costs.

The Fraunhofer IFF has determined that indirect costs account for 9–20% of the total costs. One study identified certain factors that most frequently affect indirect costs in the overall biomass supply chain:

- information and communications technology (internal and external, i.e., partners'),
- the type, duration and frequency of information exchange, and
- internal and external communication.

Clearly, factoring in transport and harvesting costs does not suffice to cut total costs.

Field tests in the Best4VarioUse project corroborated the data gathered in interviews and studies. Several different scenarios were developed for harvesting, storage, processing and transport and methods were developed to ascertain the direct and indirect costs. Among other things, morphological approaches were employed to define actions (*see* ) and select the technology to be tested.

The actions selected for the field tests were typical agricultural, forestry and landscape conservation actions, for example, the creation of networks of tending and skidding trails, which have not delivered any commercially developable yield of biomass but are necessary actions nonetheless, for example, in forestry, with potentially utilizable biomass yields.

The indirect costs, for example, for supply planning and coordination, turned out to constitute a significant portion of the biomass supply costs, but are not normally enter into costing. The smaller the biomass yields, the more considerable indirect costs. Oversized special equipment used by service providers, which is not justifiable for a cost effective cross-company supply chain, is another cost driver.

## 11.4   INTERACTION AND THE INFORMATION FLOW ARE CRUCIAL

Forestry operations, wood processing industries and the other actors involved in the chain of operations plan and control their activities based on internal, normally functional, criteria and their areas of responsibility (e.g., purchasing, logistics or marketing). The criteria are largely a function of the size of the service companies', for example, forestry service or transport providers, and their equipment and services. The different quantities ordered by and the quality requirements of differently sized consumers, from single households to medium-sized combustion plants, present another challenge.

Moreover, the many different actors are interconnected differently in different scenarios and the widest variety of client-contractor relationships exist, as the following two scenarios illustrate. In one scenario, the owner of a private forest sells the biomass from it and contracts different service providers to harvest and transport it. A consumer transports it and has a service provider chip it to the requisite quality before burning it. In another scenario, the forest owner has employees harvest and chip the biomass with a leased chipper and then has it transported to the consumer by a transport provider. Both scenarios rely on different actors and therefore, also generate different direct and, above all, indirect costs.

In addition to technical problems at the interfaces between the individual actors in the biomass supply chain, deficits in organization and IT especially impede organizing biomass value added cost effectively and ecologically sustainably. Interactions between and factors influencing the actors and their impact on the cost effectiveness and efficiency of internal and cross company operations are often overlooked.

These operations often drive indirect costs in particular. Normally, many small orders are accepted and filled in the biomass value added chain. Indirect costs are incurred by both small and large orders. Although, the costs for small orders are lower than for large orders, they increase exponentially when the total quantity ordered in both small and large orders add up to be identical. Thus, concluding larger contracts, for example, consolidating orders, can significantly cut indirect costs and thus also total costs. Many different contractors offer their services and different terms and conditions (price, type of contract and quality). The multitude of interfaces and lack of advanced information technologies often cause communication losses and thus longer wait and search times or duplicated work. Information is often gathered or transmitted incompletely and absent standards generates additional costs and cause information losses at interfaces. Therefore, good interaction among the actors and a coordinated information flow are crucial to reducing indirect costs.

## 11.4   PLANNING IS CRUCIAL

From the site of production to the consumer, the biomass supply chain is characterized by many spatially distributed, time-variable biomass sources and many spatially distributed providers and consumers (*see* Fig. 2). A multitude of actors (forest owners, forestry operations, private forestry contractors, carriers, transport companies, combustion plant operators, etc.) must collaborate in the strongly customer-oriented supply chain. This diverse and complex network of partners collaborates in different operations and exchanges information through a multitude of interfaces.

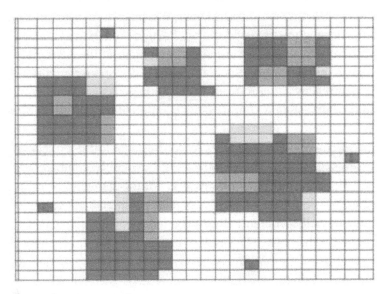

- Purchaser (1 assortment)
- Forest areas (n assortments)
- Actions of previous years
- Currently planned measures

**FIGURE 2**   Spatially distributed, time-variable biomass sources and spatially distributed providers and consumers.

Biomass's different properties (e.g., shape, weight and volume), different methods of harvesting widely varying quantities and assortments and different means of processing at individual locations constitute other specific challenges when planning operations. The quantity and composition of the biomass may vary greatly depending on the type, scope and location of the forestry, agricultural and landscape conservation actions. The quality and quantity of the biomass from landscape conservation actions or forest thinning may also vary greatly. Consumer requirements vary just as greatly. Small and medium-sized combustion plants require chipped wood measuring between 8.5 cm and 25 cm in size (length).

This raises the question: what quantity of which biomass is to be harvested with what equipment and at what cost for which consumer(s)? Thus, the focus of cost optimization shifts to strategic and operative planning. Strategic planning must include an early, integrated analysis of the site of production and the biomass consumers. Identifying the location and plant cover of lands, the infrastructure, distances to consumers and their requirements, and the time and scope of actions is essential.

Operatively, planning for the right equipment in the right place at the right time is crucial. This especially affects direct supply costs, for example, incurred by the failure to fully use expensive special equipment. In addition, harvesting and processing equipment greatly affect biomass's quality.

The aforementioned field tests in the Best4VarioUse project corroborated these statements. The practical tests explicitly compared different methods of harvesting and processing. Motor-manual harvesting with chainsaws, brush cutters or similar special equipment often proves to be superior, especially economically. Large and special equipment is often inappropriate for the material's small dimensions or the actions themselves. On the other hand, motor-manual harvesting is extremely flexible and thus expedites processing.

## 11.5   OPTIMIZING THE OVERALL OPERATION

Optimal strategic and operative planning is indispensible to optimizing operations and costs. Marketing decisions can only be made when costs and sale prices have been compared throughout the entire biomass supply chain, factoring in different scenarios and alternative consumers. Continuous modification of the equipment used ensures that the biomass is supplied as efficiently as possible. This entails determining whether special or universal equipment should be invested in or existing machinery can be modified with less expense. Information on the site of use, operating conditions and restrictions and the quality of the biomass are important for such decisions. Additionally, this information is an indispensible criterion for selecting methods. Pooling different harvesting, processing or transport equipment is an expedient cost cutting strategy.

Taking advantage of potentials for collaborations to systematically combine interfaces and capabilities is another means to optimize total operational costs. Any commercial risks need to be assessed beforehand. Ultimately, the range of services can usually be expanded. Thinking outside of the box is indispensible to taking this initial step toward a value added network. Individual operations and activities must be examined critically in order to be able to draft contracts expediently and calculate costs realistically. The correlation between one's own and one's network partners' success must always be borne in mind.

Consistent use of advanced ICT can significantly cut costs in the overall operation and is instrumental to the successful implementation of integrated planning based on sound foundations and proper tools with standardized interfaces accessible to all of the biomass supply chain's actors.

When these strategies and approaches are rigorously followed, the optimized planning and control of the biomass supply chain cuts indirect and direct costs significantly and sustainably.


## 11.6  CONCLUSION

Logistics costs include far more than just harvesting and transport costs. Optimizing the value added chain ecologically and economically entails factoring in not only the direct but also the indirect supply costs of the entire biomass supply chain. The Fraunhofer IFF in Magdeburg is currently pursuing research to formulate recommendations for actions that will minimize direct costs, among other things, by applying optimal planning and control. In addition, potentials are being identified to cut indirect costs by optimizing the operations throughout the entire biomass supply chain. This will constitute a substantial step toward optimizing the biomass value added chain ecologically and economically.

## KEYWORDS

- **Best4VarioUse project**
- **biomass supply chain**
- **forest to the factory**
- **Fraunhofer IFF**

## REFERENCES

1. Aretz A.; Hirschl B. *Biomassepotenziale in Deutschland-Übersicht maßgeblicher Studienergebnisse und Gegenüberstellung der Methoden*. DENDROM Discussion Paper No.1, March **2007.**
2. Dreier T. *Ganzheitliche Systemanalyze und Potenziale biogener Kraftstoffe*. e und M, Energie-Und-Management-Verlag-ges.: Herrsching, **2000.**
3. Dietz, H.-U.; Nick, L.; Ehrhardt, I.; Urbanke, B.; Hauck, B.
4. *Standardisierung in der Forstlogistik – Notwendigkeit, Wirkung und Chancen*. In Thees O. und Lemm R. (Eds.), Management zukunftsfähige Waldnutzung Grundlagen, Methoden und Instrumente. vdf Hochschulverlag AG: Zurich, **2009.**
5. Fritsche et al. *Stoffstrom analyze zur nachhaltigen energetischen Nutzung von Biomasse*, Darmstadt, May **2004.**
6. German Federal Ministry of the Environment, Nature Conservation and Nuclear Safety BMU: *erneuerbare Energien in Zahlen – nationale und internationale Entwicklung*, Berlin, June **2010.**
7. Kaltschmitt, M.; Merten, D. et al. *Energie aus Biomasse: grundlagen, Techniken und Verfahren*. Springer Verlag: Berlin, **2009.**
8. Kreis Siegen-wittgenstein. Landschaftsplanung in Naturschutzgebieten (NSG). www.siegen-wittgenstein.de/standard/page.sys/details/eintrag_id=790/content_id=588/349.htm. July 30, **2010.**
9. Kuhn A. und Hellingrath B. *Supply Chain Management – Optimierte Zusammenarbeit in der Wertschöpfungskette*. Springer Verlag: Berlin, **2002.**
10. Murach, D.; Knur, L. und Schultze, M. *DENDROM – Zukunftsrohstoff Dendromasse Systemische Analyze, Leitbilder und Szenarien für die nachhaltige energetische und*

*stoffliche Verwertung von Dendromasse aus Wald- und Agrarholz.* Eberswalde, November **2008,**

11. Schneider St. *Entwicklung eines (teil-)automatisierten IT-prozesses zur Erzeugung digitaler Wegedaten unter Berücksichtigung landwirtschaftlicher Aspekte.* Unpublished Diplom Thesis submitted to the Department of Business Information Systems, School of Computer Science, Otto von Guericke University Magdeburg. **2010.**

12. Schulte, C.: *logistik: wege zur Optimierung der Supply Chain.* Vahlen: munich, **2005.**

13. Schuster, K.: *Überbetriebliche Aufbringung und Vermarktung von Brennhackschnitzeln.* Landtechnische Schriften No. *179,* Ö sterreichisches Kuratiorium für Landtechnick, Vienna, **1991.**

14. Ropohl, G. *Eine Systemtheorie der Technik. Zur Grundlegung der allgemeinen Technologie.* Hanser Verlag: Munich, **1979.**

15. Holzzentralblatt. *Integrierte Logistikketten für die Holzbranche.* DRW-verlag Weinbrenner GmbH + Co. KG Holz-zentralblatt. July 15, **2005.**

16. Landesforstverwaltung, NRW. *Clusterstudie Forst and Holz.* http:,www.forst.nrw.de/nutzung/cluster/cluster.htm. **2003.**

# SECTION II: SUPRAMOLECULAR DECOMPOSITION OF THE ARALKYL HYDROPEROXIDES IN THE PRESENCE OF ET$_4$NBR

N. A. TUROVSKIJ, E. V. RAKSHA, VU. V. BERESTNEVA,
E. N. PASTERNAK, M. YU. ZUBRITSKIJ, I. A. OPEIDA,
and G. E. ZAIKOV

## 11.1 INTRODUCTION

Kinetics of the aralkyl hydroperoxides decomposition in the presence of tetraethylammonium bromide (Et$_4$NBr) has been investigated. Et$_4$NBr has been shown to reveal the catalytic properties in this reaction. The use of Et$_4$NBr leads to the decrease up to 40 kJ·mol$^{-1}$ of the hydroperoxides decomposition activation energy. The complex formation between hydroperoxides and Et$_4$NBr has been shown by the kinetic and $^1$H NMR spectroscopy methods. Thermodynamic parameters of the complex formation and kinetic parameters of complex-bonded hydroperoxides have been estimated. The model of the reactive hydroperoxide – catalyst complex structure has been proposed. Complex formation is accompanied with hydroperoxide chemical activation.

Quaternary ammonium salts *exhibit high catalytic activity in* radical-chain reactions *of hydrocarbons liquid phase oxidation by O$_2$* [1, 2]. Tetraalkylammonium halides accelerate radical decomposition of hydroperoxides [3, 4] that are primary molecular products of *hydrocarbons oxidation reaction. Reaction rate of the hydroperoxides decomposition in the presence of* quaternary ammonium salts is determined by the nature of the salt anion [4] as well as cation [5]. The highest reaction rate of the *tert*-butyl hydroperoxide and cumene hydroperoxide decomposition has been observed in the case of iodide anions as compared with bromide and chloride ones [4]. *Among* tetraalkylammonium bromides tetraethylammonium one possesses the highest reactivity in the reaction of catalytic decomposition of 1-hydroxy-cyclohexyl hydroperoxide [5] and lauroyl peroxide [6]. Benzoyl peroxide – tetraalkylammonium bromide binary systems were found to be the most efficient in the liquid-phase oxidation of isopropylbenzene *although corresponded iodides revealed* the highest reactivity in the reaction of benzoyl peroxide decomposition [1, 7, 8].

Thus, the elucidation of the reaction pathways requires taking into account many parameters and makes the investigation of other peroxide compounds reactivity a justified task. To get more information on the reaction scope and limitations, we studied aralkyl hydroperoxides reactivity in the presence of quaternary ammonium salt. The analysis of the data obtained earlier enabled us to consider tetraethylammonium bromide the *appropriate* catalyst to study aralkyl hydroperoxides reactivity.

The present paper reports on new results of kinetic investigations of aralkyl hydroperoxides catalytic decomposition in the presence of tetraethylammonium bromide as well as the results of AM1/COSMO molecular modeling of hydroperoxide – catalyst interactions. Aralkyl hydroperoxides under consideration are: dimethylbenzylmethyl hydroperoxide $(PhCH_2C(CH_3)_2OOH)$, 1,1-dimethyl-3-phenylpropyl hydroperoxide $(Ph(CH_2)_2C(CH_3)_2OOH)$, 1,1-dimethyl-3-phenylbutyl hydroperoxide $(PhCH(CH_3)CH_2C(CH_3)_2OOH)$, 1,1,3-trimethyl-3-($p$-methylphenyl)butyl hydroperoxide $(p\text{-}CH_3PhC(CH_3)_2CH_2C(CH_3)_2OOH)$, $p$-carboxyisopropylbenzene hydroperoxide $((p\text{-}COOH)\text{-}phC(CH_3)_2OOH)$. *Tert*-butyl hydroperoxide $((CH_3)_3COOH)$ and cumene hydroperoxide $(Ph(CH_3)_2COOH)$ were also used in spectroscopic investigations.

## 11.2   EXPERIMENTAL METHODS

Aralkyl hydroperoxides (ROOH) were purified according to Ref. [9]. Their purity (98.9%) was controlled by iodometry method. Tetraalkylammonium bromide $(Et_4NBr)$ was recrystallized from acetonitrile solution by addition of diethyl ether excess. The salt purity (99.6%) was determined by argentum metric titration with potentiometric fixation of the equivalent point. Tetraalkyl ammonium bromide was stored in box dried with $P_2O_5$. Acetonitrile $(CH_3CN)$ was purified according to Ref. [10]. Its purity was controlled by electro conductivity $\chi$ value, which was within $(8.5 \pm 0.2) \times 10^{-6}$ $W^{-1}$ $cm^{-1}$ at 303 K.

Reactions of the hydroperoxides catalytic decomposition were carried out in glass-soldered ampoules in argon atmosphere. To control the proceeding of hydroperoxides thermolysis and their decomposition in the presence of $Et_4NBr$ the iodometric titration with potentiometric fixation of the equivalent point was used.

$^1H$ NMR spectroscopy investigations of the aralkyl hydroperoxides and hydroperoxide – $Et_4NBr$ solutions were carried out at equimolar components ratio ([ROOH] = [$Et_4NBr$] = 0.1 mol $dm^{-3}$) in $D_3CCN$ at 294 K. The $^1H$ NMR spectra were recorded on a Bruker Avance 400 (400 MHz) using TMS as an internal standard.

*Dimethylbenzylmethyl hydroperoxide* ($PhCH_2C(CH_3)_2OOH$) $^1H$ NMR (400 MHz, acetonitrile-$d_3$): $\delta$ = 1.11 (s, 6 H, $CH_3$), 2.84 (s, 2 H, $CH_2$), 7.20–7.31 (m, 5H, H-aryl), 8.88 (s, 1 H, -COOH) ppm.

*1,1-dimethyl-3-phenylpropyl hydroperoxide* ($Ph(CH_2)_2C(CH_3)_2OOH$) $^1H$ NMR (400 MHz, acetonitrile-$d_3$): $\delta$ = 1.22 (s, 6 H, $CH_3$), 1.80 (t, $J$ = 8.0 Hz, 2 H, Ph-$CH_2$ $CH_2$), 2.64 (t, $J$ = 8.0 Hz, 2 H, Ph-$CH_2$ $CH_2$), 7.15–7.30 (m, 5 H, H-aryl), 8.86 (s, 1 H, -COOH) ppm.

*1,1,3-trimethyl-3-(p-methylphenyl)butyl hydroperoxide* (p-$CH_3PhC(CH_3)_2CH_2C(CH_3)_2OOH$) $^1H$ NMR (400 MHz, acetonitrile-$d_3$): $\delta$ = 0.86 (s, 6 H, -$C(CH_3)_2OOH$), 1.34 (s, 6 H, -ph$C(CH_3)_2$), 2.03 (s, 2 H, -$CH_2$), 2.29 (s, 3 H, $CH_{3-p}$h-), 7.10 (d, $J$ = 8.0 Hz, 2 H, H-aryl), 7.30 (d, $J$ = 8.0 Hz, 2 H, H-aryl), 8.51 (s, 1 H, -COOH) ppm.

*p-carboxyisopropylbenzene hydroperoxide* ((p-COOH)$PhC(CH_3)_2OOH$) $^1H$ NMR (400 MHz, acetonitrile-$d_3$): $\delta$ = 1.53 (s, 6 H, -$C(CH_3)_2OOH$), 7.57 (d, $J$ = 8.0 Hz, 2 H, H-aryl), 7.98 (d, $J$ = 8.0 Hz, 2 H, H-aryl) ppm. Signals from -COOH and -C(O)OH exchaneable protons were not observed.

*Tert-butyl hydroperoxide* (($CH_3)_3COOH$) $^1H$ NMR (400 MHz, acetonitrile-$d_3$): $\delta$ = 1.18 (s, 6 H, –$CH_3$), 8.80 (s, 1 H, -COOH) ppm.

*Cumene hydroperoxide* ($Ph(CH_3)_2COOH$) $^1H$ NMR (400 MHz, acetonitrile-$d_3$): $\delta$ = 1.52 (s, 6 H, -$CH_3$), 7.27 (t, $J$ = 8.0 Hz, 1 H, H-aryl), 7.36 (t, $J$ = 8.0 Hz, 2 H, H-aryl), 7.47 (d, $J$ = 8.0 Hz, 2 H, H-aryl), 8.95 (s, 1 H, -COOH) ppm.

*Tetraethylammonium bromide* (Et$_4$NBr) $^1$H NMR (400 MHz, acetoni-trile-d$_3$): $\delta$ = 1.21 (t, $J$ = 8.0 Hz, 12 H, –CH$_3$), 3.22 (q, $J$ = 8.0 Hz, 8 H, -CH$_2$) ppm.

Quantum chemical calculations of the equilibrium structures of hydroperoxides molecules and corresponding radicals as well as tetraethylammonium bromide and ROOH – Et$_4$NBr complexes were carried out by AM1 semiempirical method implemented in MOPAC2009™ package [11]. The RHF method was applied to the calculation of the wave function. Optimization of hydroperoxides structure parameters was carried out by Eigenvector following procedure. The molecular geometry parameters were calculated with boundary gradient norm 0.01. The nature of the stationary points obtained was verified by calculating the vibrational frequencies at the same level of theory. Solvent effect in calculations was considered in the COSMO [12] approximation.

## 11.3   RESULTS AND DISCUSION

**Kinetics of the aralkyl hydroperoxides decomposition in the presence of the Et$_4$NBr.** Kinetics of the aralkyl hydroperoxides decomposition in the presence of Et$_4$NBr has been studied under conditions of ammonium salts excess in the reaction mixture. Reactions were carried out at 373–393 K, hydroperoxide initial concentration was $5 \times 10^{-3}$ mol dm$^{-3}$, Et$_4$NBr concentration in the system was varied within $2 \times 10^{-2} - 1.2 \times 10^{-1}$ mol dm$^{-3}$. Hydroperoxides decomposition kinetics could be described as the first order one. The reaction was carried out up to 80% hydroperoxide conversion and the products did not affect the reaction proceeding, as the kinetic curves anamorphous were linear in the corresponding first order coordinates.

The reaction effective rate constant ($k_{ef}$, s$^{-1}$) was found to be independent on the hydroperoxide initial concentration within [ROOH]$_0$ = $1 \times 10^{-3}$ – $8 \times 10^{-3}$ mol dm$^{-3}$ at 383 K, however, Et$_4$NBr amount was kept constant ($5 \times 10^{-2}$ mol dm$^{-3}$) in these studies. This fact allows one to exclude the simultaneous hydroperoxides reactions in the system under consideration.

Typical nonlinear $k_{ef}$ dependences on Et$_4$NBr concentration at the constant hydroperoxide initial concentration are presented on Fig. 1a. The nonlinear character of these dependences allows us to assume that an intermediate adduct between ROOH and $Et_4NBr$ is formed in the reaction of ROOH decomposition in the presence of $Et_4NBr$. These facts conform to the kinetic scheme of activated cumene hydroperoxide and hydroxycyclohexyl hydroperoxide decomposition that has been proposed *previously* [3–5, 13].

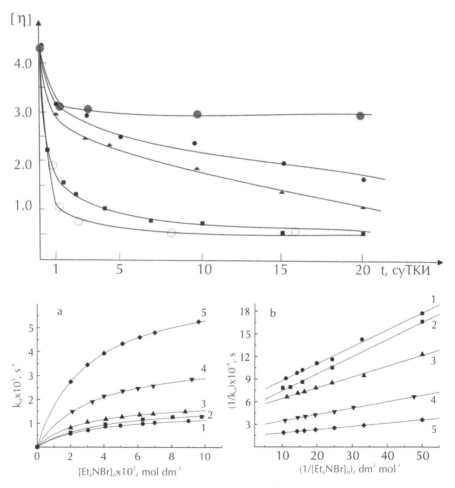

**FIGURE 1**   Dependence of $k_{ef}$ on the $Et_4NBr$ initial concentration in direct (a) and inverse (b) coordinates ($[ROOH]_0 = 5.0 \times 10^{-3}$ mol dm$^{-3}$, 383 K).
*1* – $PhCH_2C(CH_3)_2OOH$; *2* – $Ph(CH_2)_2C(CH_3)_2OOH$; *3* – $PhCH(CH_3)CH_2C(CH_3)_2OOH$; *4* – $p$-$CH_3PhC(CH_3)_2CH_2C(CH_3)_2OOH$; *5* – $p$-$HOC(O)PhC(CH_3)_2OOH$

The hydroperoxide – Et$_4$NBr kinetic mixtures were subjected to $^1$H NMR analysis in order to confirm the complex formation between ROOH and Et$_4$NBr. Experiment was carried out at 298 K when the rates of the hydroperoxides thermolysis as well as activated decomposition were negligibly small. Fig. 2a presents $^1$H NMR spectra of the p-CH$_3$PhC(CH$_3$)$_2$CH$_2$C(CH$_3$)$_2$OOH in CD$_3$CN. Chemical shift of the hydroperoxide moiety of the compound corresponds to the signal at 8.51 ppm. Addition of the Et$_4$NBr equivalent amount to the solution causes the dislocation of this signal by 0.51 ppm towards weak *magnetic fields (Fig. 2b)*.

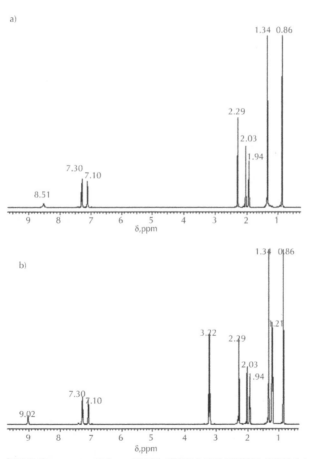

**FIGURE 2**   $^1$H NMR spectra of the p-CH$_3$PhC(CH$_3$)$_2$CH$_2$C(CH$_3$)$_2$OOH (*a*) and mixture of the p-CH$_3$PhC(CH$_3$)$_2$CH$_2$C(CH$_3$)$_2$OOH – Et$_4$NBr (*b*) in CD$_3$CN at 298 K [ROOH] = [Et$_4$NBr] = 1.0×10$^{-1}$ mol dm$^{-3}$.

*The similar effect was also observed for the rest investigated hydroper-oxides. Thus* presence of the $Et_4NBr$ in the solution causes the downfield shifts of the COOH moiety signal by 0.40–0.76 ppm as compared to the chemical shift of the hydroperoxide (Table 1) depending on the hydroper-oxide structure. This effect is typical for the hydroperoxide complexation processes.

**TABLE 1**    $^1H$ NMR spectra parameters of the hydroperoxides and the hydroperoxide – $Et_4NBr$ mixture in $D_3CCN$ at 294 K.

| ROOH | δ, ppm (ROO–H) | | |
|------|------|------|------|
| | **ROOH** | **ROOH + $Et_4NBr$** | **Δδ** |
| $PhCH_2C(CH_3)_2OOH$ | 8.88 | 9.64 | 0.76 |
| $p\text{-}CH_3PhC(CH_3)_2CH_2C(CH_3)_2OOH$ | 8.51 | 9.02 | 0.51 |
| $(CH_3)_3COOH$ | 8.80 | 9.20 | 0.40 |
| $Ph(CH_3)_2COOH$ | 8.95 | 9.60 | 0.65 |

Chemically activated by $Et_4NBr$ aralkyl hydroperoxides decomposi-tion is suggested to proceed in accordance with following kinetic scheme (1). It includes the stage of a complex formation between the hydroperox-ide molecule and $Et_4NBr$ ions as well as the stage of complex – bonded hydroperoxide decomposition.

$$ROOH + Et_4NBr \xrightleftharpoons{K_C} [complex] \xrightarrow{k_d} products \qquad (1)$$

where $K_C$ – *equilibrium* constant of the complex formation ($dm^3\ mol^{-1}$), and $k_d$ – rate constant of the complex decomposition ($s^{-1}$).

As the kinetic parameters of hydroperoxides decomposition are in-fluenced by the nature of the salt anion [4] as well as cation [5, 6] we assume the model of $ROOH\text{-}Et_4NBr$ complex formation with combined action of cation and anion. The hydroperoxides thermolysis *contribution* to the overall reaction rate of the ROOH activated decomposition is neg-ligibly small because thermolysis rate constants [14] are less by an order of magnitude then correspondent $k_{ef}$ values. Using a kinetic model for the generation of active species (Scheme 1) and analyzing this scheme in a quasi-equilibrium approximation one can obtain the following equation for the $k_{ef}$ dependence on $Et_4NBr$ concentration:

$$k_{ef} = \frac{K_C k_d [Et_4NBr]_0}{1 + K_C [Et_4NBr]_0} \qquad (2)$$

To simplify further analysis of the data from Fig. 1a, let us transform Eq. (2) into the following one:

$$\frac{1}{k_{ef}} = \frac{1}{k_d K_C [Et_4NBr]_0} + \frac{1}{k_d} \qquad (3)$$

Eq. (3) can be considered as equation of straight line in $(\frac{1}{k_{ef}} - \frac{1}{[Et_4NBr]_0})$ coordinates. The $k_{ef}$ dependences on $Et_4NBr$ concentration are linear in double inverse coordinates (Fig. 1b). Thus the experimentally obtained parameters with reasonable accuracy correspond to the proposed kinetic model and are in the quantitative agreement with this model if we assume that $K_C$ and $k_d$ have correspondent values listed in Table 2.

*Values of equilibrium constants of* complex formation $(K_C)$ between ROOH and $Et_4NBr$ estimated are within 21–34 $dm^3$ $mol^{-1}$ (at 273–293 K) for the investigated systems. It should be noted that $k_d$ values do not depend on the ROOH and $Et_4NBr$ concentration and correspond to the *ultimate case when all hydroperoxide molecules are complex-bonded and further addition of* $Et_4NBr$ to the reaction mixture will not lead to the increase of the reaction rate.

Estimated values of complex formation reaction enthalpies $(\Delta H_{com}$ in Table 2) are within $(-15 \div -22)$ $kJ$ $mol^{-1}$ and correspond to the hydrogen bond energy in weak interactions [15]. Considering the intermolecular bonds energy, the strongest complex is formed between $Et_4NBr$ and hydroperoxide $PhCH(CH_3)CH_2C(CH_3)_2OOH$ (*see* the corresponding $\Delta H_{com}$ values in Table 2).

Symbate changes in thermolysis and catalytic decomposition activation energies are observed for considered aralkyl hydroperoxides (Fig. 3). Thus peroxide bond *cleavage causes the activation energy of the complex-bonded hydroperoxide decomposition.*

Hydroperoxide $p$-HO(O)CPhC(CH$_3$)$_2$OOH in which hydroperoxide moiety is directly connected with aromatic ring shows the highest reactivity. Aliphatic group *occurrence between* hydroperoxide moiety and aromatic ring leads to the *decrease of* hydroperoxide reactivity in the reaction of catalytic decomposition in the presence of Et$_4$NBr. According to the $E_a$ values listed in Table 2 the reactivity of the complex-bonded hydroperoxides increases as follows: phCH$_2$C(CH$_3$)$_2$OOH < Ph(CH$_2$)$_2$C(CH$_3$)$_2$OOH < PhCH(CH$_3$)CH$_2$C(CH$_3$)$_2$OOH < ($p$-CH$_3$)PhC(CH$_3$)$_2$CH$_2$C(CH$_3$)$_2$OOH < $p$-HO(O)CPhC(CH$_3$)$_2$OOH. Using of Et$_4$NBr allows decreasing by 40 kJ mol$^{-1}$ the activation energy of hydroperoxides decomposition in acetonitrile.

**TABLE 2**  Kinetic parameters of aralkyl hydroperoxides decomposition activated by Et$_4$NBr.

| ROOH[1] | $T$ (K) | $k_d \times 10^5$ (s$^{-1}$) | $K_C$ (dm$^3$ mol$^{-1}$) | $E_a$ (kJ mol$^{-1}$) | $\Delta H_{com}$ (kJ mol$^{-1}$) |
|---|---|---|---|---|---|
| 1 | 373 | 0.89 ± 0.04 | 27 ± 1 | 100 ± 3 | −15 ± 1 |
|   | 383 | 1.94 ± 0.09 | 23 ± 2 | | |
|   | 393 | 4.6 ± 0.2 | 21 ± 2 | | |
| 2 | 373 | 1.52 ± 0.09 | 29 ± 3 | 96 ± 5 | −17 ± 2 |
|   | 383 | 3.75 ± 0.08 | 26 ± 1 | | |
|   | 393 | 7.3 ± 0.1 | 22 ± 1 | | |
| 3 | 373 | 2.00 ± 0.06 | 34 ± 2 | 92 ± 4 | −22 ± 2 |
|   | 383 | 4.3 ± 0.1 | 27 ± 3 | | |
|   | 393 | 8.5 ± 0.1 | 24 ± 2 | | |
| 4 | 373 | 3.85 ± 0.07 | 31 ± 3 | 88 ± 5 | −19 ± 2 |
|   | 383 | 9.07 ±0.09 | 26 ± 2 | | |
|   | 393 | 17.3 ± 0.6 | 23 ± 2 | | |
| 5 | 373 | 3.32 ± 0.06 | 37 ± 2 | 85 ± 1 | −15 ± 1 |
|   | 383 | 6.90 ± 0.05 | 33 ± 3 | | |
|   | 393 | 13.4 ± 0.4 | 29 ± 3 | | |

$^1$ROOH: 1 – PhCH$_2$C(CH$_3$)$_2$OOH; 2 – Ph(CH$_2$)$_2$C(CH$_3$)$_2$OOH; 3 – PhCH(CH$_3$)CH$_2$C(CH$_3$)$_2$OOH; 4 – ($p$-CH$_3$)PhC(CH$_3$)$_2$CH$_2$C(CH$_3$)$_2$OOH; 5 – $p$-HO(O)CPhC(CH$_3$)$_2$OOH

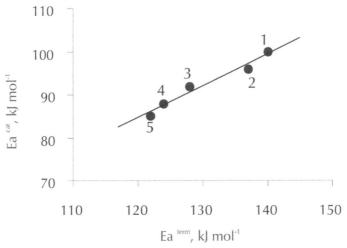

**FIGURE 3**   Symbate changes in thermolysis ($E_a{}^{term}$) and catalytic ($E_a{}^{cat}$) decomposition activation energies of aralkyl hydroperoxides ($E_a{}^{term}$ values are listed elsewhere [14]).

**Molecular modeling of the Et$_4$NBr activated hydroperoxides decomposition.** On the base of experimental facts mentioned above we consider the salt anion and cation as well as acetonitrile (solvent) molecule participation when model the possible structure of the reactive hydroperoxide – catalyst complex. Model of the substrate separated ion pair (Sub-SIP) is one of the possible realization of join action of the salt anion and cation in hydroperoxide molecule activation. In this complex hydroperoxide molecule is located between cation and anion species. For the symmetric molecules such as benzoyl peroxide [16], lauroyl peroxide [17], and dihydroxydicyclohexyl peroxide [18] attack of salts ions is proposed to be along direction of the peroxide dipole moment and perpendicularly to the peroxide bond. Hydroperoxides are asymmetric systems, that is why different directions of ion's attack are possible. The solvent effect can be considered by means of direct inclusion of the solvent molecule to the complex structure. From the other hand methods of modern computer chemistry allow estimation the solvent effect in continuum solvation models approximations [12, 19, 20].

Catalytic activity of the $Et_4NBr$ in the reaction of the aralkyl hydroperoxides decomposition can be considered due to chemical activation of the hydroperoxide molecule in the salt presence. Activation of a molecule is the modification of its electronic and nucleus structure that leads to the increase of the molecule reactivity. The SubSIP structural model has been obtained for the complex between hydroperoxide molecule and $Et_4NBr$ (Fig. 4).

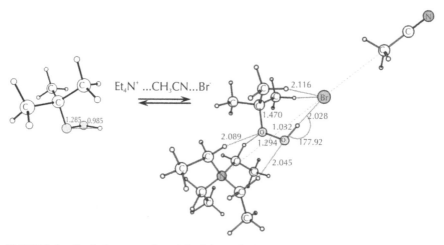

**FIGURE 4**   Typical structural model of the aralkyl hydroperoxides – $Et_4NBr$ complexes with combined action of the cation, anion and the solvent molecule obtained by AM1/ COSMO method.

Investigation of the SubSIP structural model properties for considered aralkyl hydroperoxides has revealed that complex formation was accompanied by following structural effects: (i) peroxide bond elongation on 0.02 Å as compared with nonbonded hydroperoxide molecules; (ii) considerable conformation changes of the hydroperoxide fragment; (iii) O–H bond elongation on 0.07 Å in complex as compared with nonbonded hydroperoxide molecules; (iv) rearrangement of electron density on the hydroperoxide group atoms.

Optimization of complex bonded hydroperoxides in COSMO approximation has shown that solvent did not effect on the character of structural changes in hydroperoxides molecules in the SubSIP complex. Only partial electron density transfer from bromide anion on to peroxide bond is less noticeable in this case.

Hydroperoxide moiety takes part in the formation of hydrogen bond (O)H...Br- in all obtained complexes. In all structures distances (O)H...Br- are within $2.03 \div 2.16$ Å, bond angle O-H...Br- is higher than $90°$ and within $170° \div 180°$ corresponding from aralkyl moiety in the hydroperoxide molecule. Thus the interaction type of bromide anion with hydroperoxide can be considered as hydrogen bond [15].

Associative interactions of a hydroperoxide molecule with $Et_4NBr$ to the peroxide bond dissociation energy ($D_{O-O}$) decrease. $D_{O-O}$ value for the aralkyl hydroperoxide was calculated according to Eq. (4) and for the $ROOH - Et_4NBr$ complexes – according to Eq. (5).

$$D_{O-O} = (\Delta_f H°(RO^\cdot) + \Delta_f H°(^\cdot OH)) - \Delta_f H°(ROOH), \quad (4)$$

$$D_{O-O} = (\Delta_f H°(RO^\cdot) + \Delta_f H°(^\cdot OH)) - \Delta_f H°(ROOH_{comp}), \quad (5)$$

where $\Delta_f H°(RO^\cdot)$ – standard heat of formation of the corresponding oxi-radical; $\Delta_f H°(^\cdot OH)$ – standard heat of formation of the $^\cdot OH$ radical; $\Delta_f H°(ROOH)$ – standard heat of formation of the corresponding hydroperoxide molecule; $\Delta_f H°(ROOH_{comp})$ – heat of formation of the hydroperoxide molecule that corresponds to the complex configuration. $D_{O-O}$ value for the hydroperoxide configuration that corresponds to complex one is less than $D_{O-O}$ for the nonbonded hydroperoxide molecule. Difference between bonded and nonbonded hydroperoxide $D_{O-O}$ value is $\Delta D_{O-O} = (43 \pm 5)$ kJ·mol$^{-1}$ (Table 3) and in accordance with experimental activation barrier decreasing in the presence of $Et_4NBr$: $\Delta E_a = (40 \pm 3)$ kJ·mol$^{-1}$.

Linear dependence has been obtained between experimental activation energy of the aralkyl hydroperoxide – $Et_4NBr$ complex decomposition and calculated value of $\Delta D_{O-O}$ that characterizes the peroxide bond strength decreasing (Fig. 5a). Thus changes in the hydroperoxide moiety configuration during complex formation lead to the destabilization of the peroxide bond, its strength decreasing, and to the increasing of the hydroperoxide molecule reactivity.

**TABLE 3** Values of $\Delta_f H°(ROOH)$, $\Delta_f H°(ROOH_{comp})$, $\Delta D_{O-O}$ i $\Delta_r H°$ for the aralkyl hydropero-xides obtained with AM1/COSMO method.

| [1]ROOH | $\Delta_f H°(ROOH)$, kJ·mol$^{-1}$ | $\Delta_f H°(ROOH_{comp})$, kJ·mol$^{-1}$ | $\Delta D_{O-O}$, kJ·mol$^{-1}$ | $\Delta_r H°$, kJ·mol$^{-1}$ |
|---|---|---|---|---|
| 1 | −69.3 | −31.2 | 38.1 | 68.8 |
| 2 | −99.0 | −59.7 | 39.1 | 65.9 |
| 3 | −107.9 | −66.9 | 41.0 | 62.7 |
| 4 | −128.7 | −84.7 | 44.0 | 58.4 |
| 5 | −409.4 | −360.9 | 48.5 | 57.5 |
| 6 | −34.3 | 13.0 | 47.3 | 56.7 |

[1]ROOH: 1 – PhCH$_2$C(CH$_3$)$_2$OOH; 2 – Ph(CH$_2$)$_2$C(CH$_3$)$_2$OOH; 3 – PhCH(CH$_3$)CH$_2$-C(CH$_3$)$_2$OOH; 4 – (p-CH$_3$)PhC(CH$_3$)$_2$CH$_2$C(CH$_3$)$_2$OOH; 5 – p-HO(O)CPhC(CH$_3$)$_2$OOH; 6 – Ph(CH$_3$)$_2$COOH.

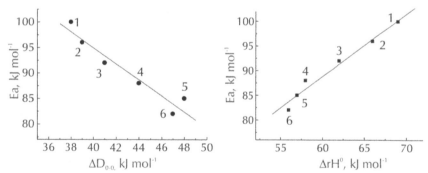

**FIGURE 5** Dependence between experimental $E_a$ of the aralkyl hydroperoxides decomposition in the presence of Et$_4$NBr and (a) – calculated $\Delta D_{O-O}$ values; (b) – calculated $\Delta_r H°$ values for the reaction (6).

Complex formation with structure of substrate separated ion pair is the exothermal process. Part of the revealed energy can be spent on structural reorganization of the hydroperoxide molecule (structural changes in –COOH group configuration). It leads to the corresponding electron reorganization of the reaction center (peroxide bond). Thus the increase of the hydroperoxide reactivity occurs after complex-bonding of the hydroperoxide molecule. So the chemical activation of the hydroperoxide molecule is observed as a result of the hydroperoxide interaction with Et-$_4$NBr. This activation promotes radical decomposition reaction to proceed in mild conditions.

In the framework of proposed structural model the hydroperoxide molecule is directly bonded in complex with ammonium salt anion and cation. This fact is in accordance with experimental observation of the anion and cation nature effects on the kinetic parameters of the activated hydroperoxide decomposition. Solvated anion approximation allows to directly account the solvent effect on the reactivity of complex bonded aralkyl hydroperoxides.

## 11.4  CONCLUSIONS

Investigations of kinetics of the aralkyl hydroperoxides decomposition in the presence of Et$_4$NBr have revealed that reaction occurred through the complex formation stage. The complex formation enthalpy value is within $(-10 \div -22)$ kJ·mol$^{-1}$ corresponding from aralkyl substituent in the hydroperoxide structure. Et$_4$NBr addition leads to the hydroperoxide decomposition activation energy decrease on 40 kJ·mol$^{-1}$. The hydroperoxides reactivity increases in the following way: $phCH_2C(CH_3)_2OOH < Ph(CH_2)_2C(CH_3)_2OOH < Ph(CH_3)CHCH_2C(CH_3)_2OOH < (p\text{-}CH_3)Ph(CH_3)_2CCH_2C(CH_3)_2OOH < p\text{-}HO(O)CPhC(CH_3)_2OOH$. The structural model of reactive complex was proposed that allowed to account the hydroperoxide nature as well as ammonium salt anion and cation effect, and the solvent one. Formation of the complex with proposed structural features is accompanied with chemical activation of the aralkyl hydroperoxide molecule.

## KEYWORDS

- acetonitrile
- aralkyl hydroperoxides
- peroxide bond
- tetraethylammonium bromide

## REFERENCES

1. Opeida, I. A.; Zalevskaya, N. M.; Turovskaya, E. N. Oxidation of Cumene in the Presence of the Benzoyl Peroxide – Tetraalkylammonium Iodide Initiator System. *Petroleum Chemistry*, **2004**, *44,* 328.

2. Matienko, L. I.; Mosolova, L. A.; Zaikov, G. E. Selective catalytic oxidation of hydrocarbons. New prospects. *Russ Chem Rev*, **2009**, *78,* 221.

3. Turovskyj, M. A.; Nikolayevskyj, A. M. ; Opeida, I. A. N. Shufletuk: Cumene hydroperoxide decomposition in the presence of tetraethylammonium bromide. *Ukrainian Chem. Bull.;* **2000**, 8, 151.

4. Turovskij, N. A.; Antonovsky, L.; Opeida, I. A.; Nikolayevskyj, A. M.; Shufletuk, N. Effect of the onium salts on the cumene hydroperoxide decomposition kinetics. russian J. *Physical Chemistry B.;* **2001**, *20,* 41.

5. Turovskij, N. A.; Raksha, E.; Gevus, O. I.; Opeida, I. A.; Zaikov, G. E. Activation of 1-hydroxycyclohexyl Hydroperoxide Decomposition in the Presence of $Alk_4NBr$. *Oxidation Communications,* **2009**, *32,* 69.

6. Turovskij, N. A.; Pasternak, E. N.; Raksha, E.; Golubitskaya, N. A.; Opeida, I. A.; Zaikov, G. E. Supramolecular Reaction of Lauroyl Peroxide with Tetraalkylammonium Bromides. *Oxidation Communications,* **2010**, *33,* 485.

7. I. Opeida, A. N.; Zalevskaya, M. E.; Turovskaya, N.; Yi. U.; Sobka: Oxidation of Cumene with Oxygen in the Presence of the Benzoyl Peroxide–Tetraalkylammonium Bromide Low-Temperature Initiator System. Petroleum Chemistry, **2002**, *42,* 423.

8. Opeida, I. A.; Zalevskaya, N. M.; Turovskaya, E. N. Benzoyl Peroxide–Tetraalkylammonium Iodide System As an Initiator of the Low-Temperature Oxidation of Cumene. Kinetics And Catalysis, **2004**, *45,* 774.

9. Hock, H.; Lang, S. Autoxidation of hydrocarbons (VIII) Octahydroanthracene peroxide. *Chem. Ber.;* **1944**, *77,* 257.

10. Vaisberger, A.; Proskauer, E.; Ruddik, J.; Tups, E. Organic Solvents. [Russian Translation]. Moscow: Izd. Inostr. Lit.; **1958.**

11. Stewart, J. J. MOPAC2009, Stewart Computational Chemistry, Colorado Springs, Colorado, USA, http:,OpenMOPAC.net.

12. Klamt Cosmo, A. A New Approach to Dielectric Screening in Solvents with Explicit Expressions for the Screening Energy and its Gradient. Chem j, soc. Perkin trans-.; 2, 799 **1993,**

13. Turovskyj, M. A.; Opeida, I. O.; Turovskaya, O. M.; Raksha, O.; Kuznetsova, N. O.; Zaikov, G. E. Kinetics of radical chain cumene oxidation initiated by α-oxycyclohexylperoxides in the presence of Et4NBr. *Oxidation Communications,* **2006**, *29,* 249.

14. Turovskyj, M. A.; Opeida, I. O.; Turovskaya, O. M.; Raksha, O. Molecular modeling of aralkyl hydroperoxides homolysis. *Oxidation Communications, 30,* 504 **2007,**

15. Williams, D. H.; Westwell, M. S. Aspects of week interactions. *Chesoc M, Rev.;* **1998,** *28,* 57.

16. M. A. Turovskyj, Yu, S. Tselinskij: Quantum chemical analysis of diacyl peroxides decomposition activated by quaternary ammonium chloride salts. *Ukrainian Chem. Bull.;* **1994,** *60,* 16

17.  Turovskij, N. A.; Pasternak, E. N.; Raksha, E.; Opeida, I. A.; Zaikov, G. E. Supra-molecular decomposition of lauroyl peroxide activated by tetraalkylammonium bromides. In: *Success in Chemistry and Biochemistry: Mind's Flight in Time and Space.* 4. – Howell New York: Nova Scince Publishers, Inc. **2009,** 555–573.

18.  Turovskyj, M. A.; Opeida, I. O.; Turovskaya, O. M.; Raksha, O.; Kuznetsova, N. O.; Zaikov, G. E. Kinetics of Activated by Et4NBr Alfa-oxycyclohexylperoxides Decomposition: Supramolecular Model. In Order and Disorder in Polymer Reactivity. Edited by G. E. Zaikov and Howell, B. A.; New York: Nova Science Publishers, Inc. **2006,** 37–51.

19.  Mennucci, B.; Tomasi J. A new approach to the problem of solute's charge distribution and cavity boundaries. *Chem J, Phys.;* **1997,** *106,* 5151.

20.  Cossi, M.; Scalmani, G. Rega, Barone, N. New developments in the polarizable continuum model for quantum mechanical and classical calculations on molecules in solution. *Chem J, Phys.;* **2002,** *117,* 43.

# SECTION III: THE USE OF THE METHOD OF DETERMINING REDUCING CARBOHYDRATES IN CONNECTION WITH THE QUESTION OF CHITOSAN ENZYMATIC DESTRUCTION

E. I. KULISH, V. V. CHERNOVA, V. P. VOLODINA,
S. V. KOLESOV, and G. E. ZAIKOV

The possibility of using viscosimetry methods and the determination of reducing carbohydrates concentration in investigation of chitosan enzymatic destruction have been discussed. It has been demonstrated that the change of chitosan intrinsic viscosity can result both from the rearrangement of the polymer supermolecular structure and from its hydrolysis. In this connection the direct method of determination of the reducing carbohydrates concentration seems to be much more correct at investigating the chitosan enzymatic destruction.

## 11.1   INTRODUCTION

The problem of chitosan (CHT) hydrolysis, including enzymatic one, has been paid sufficiently much attention to in the recent years [1–6]. Many authors [4–6] choose the viscosimetric method as that of evaluating the change in the polymer molecular mass during hydrolysis. The method of viscosimetric evaluation of the molecular mass decrease makes it possible to establish the fact of the CHT hydrolysis process by the decrease in the intrinsic viscosity of the solutions. Meanwhile, the change in the polymer solution viscosity can refect not only the change in the polymer molecular weight but also the possible rearrangement of the solution structural condition. In this connection, the aim of this investigation is the evaluation of correctness of using viscosimetry method in the studies on CHT enzymatic destruction in solution in the presence of some enzymes nonspecific for CHT.

## 11.2   EXPERIMENTAL METHODS

The objects of investigation were: food chitosan produced by the company "Bioprogress" (Schelkovo) which was obtained by alkaline deacetylation of crab chitin (the deacetylation degree is ~83%, the molecular mass is 87 kDa) and enzymatic preparations: food collegenase ("Bioprogress", Schelkovo) food pepsin ("Shako", Rostov-on-don), lidase ("Immunopreparat", Ufa), crystalline tripsin ("Microgen", Omsk). The chitosan solution with the concentration of 2% mass was prepared by dissolving during 24 hours at room temperature. Acetic acid with the concentration of 1 g/dl was used as the solvent. The enzymatic preparation previously dissolved in a small quantity of water was introduced into the polymer solution in amount of 5% of chitosan mass. Enzymatic destruction was carried out at 25°C for 10–180 hours. After exposure with enzyme the polymer was extracted from the solution, washed with distilled water and dried up to its constant weight. To determine the CHT intrinsic viscosity the 0.3% polymer solution in acetate buffer with pH=4.5 was prepared. The definition of the reducing carbohydrates concentration was carried out by the ferricyanide method [7].

## 11.3   THE RESULTS DISCUSSION

During the investigation conducted by the authors two facts have been established which seem to be mutually eliminating. On the one hand, the standing of CHT solutions with enzymes nonspecific for it – trypsin, lidase, collagenase and pepsin is accompanied by the decrease in the value of the intrinsic viscosity of its solutions (Fig. 1, curves 1–4). This fact was explained in terms of the process of CHT main chain destruction taking place under the action of enzymes. On the other hand, even if enzyme is absent CHT dissolving in acetic acid is also accompanied by a sharp one-fold change (decrease) in the value of its intrinsic viscosity. So, the value of intrinsic viscosity of CHT which didn't undergo the stage of dissolving in acetic acid, determined in acetate buffer is [η] = 4.2 dl/g. If we hold CHT in 1% acetic acid during 2–500 hours (Fig. 1, curve 5), extract it from the solution and determine its intrinsic viscosity in acetate buffer the obtained value [η] is 3.1 dl/g, however, it doesn't depend on the time of standing of the initial polymer solution. The fact of CHT viscosity decrease in acetic acid solutions has been discussed in the literature and explained in terms of rearrangement of its supermolecular structure [5, 10]. Really, the change in the polymer solution viscosity can reflect not only the change in the polymer molecular weight but also a possible rearrangement in the solution of macromolecules aggregates.

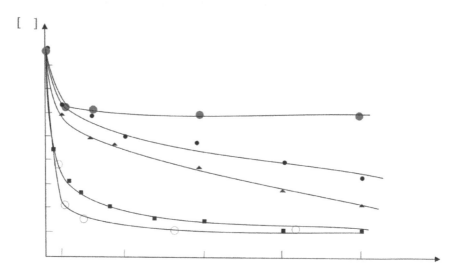

**FIGURE 1**   The dependence of the intrinsic viscosity of chitosan extracted from 1% acetic acid solution containing trypsin (1), lidase (2), collagenase (3) and pepsin (4) on the time of chitosan standing in solution.

Thus, one and the same experimental fact – the decrease in the intrinsic viscosity of CHT specimens obtained under similar experimental conclusions can be considerable in these two cases in different ways – in the presence of enzymes one can speak about the hydrolysis process, in the absence of enzymes about structural rearrangements in solution. Thus, the conclusion about CHT destruction taking place under the action of non-specific enzymes can be ambiguous if made according to the viscosimetry data. The dependences $[\eta] = f(t)$ can be in favor of destruction. The intrinsic viscosity is seen to decrease regularly with the increase of the time of exposure of the polymer – enzyme system.

According to the scheme of the hydrolytic rupture of the glycoside bond (Fig. 2) the resulting destruction products are carbohydrates having reducing capacity.

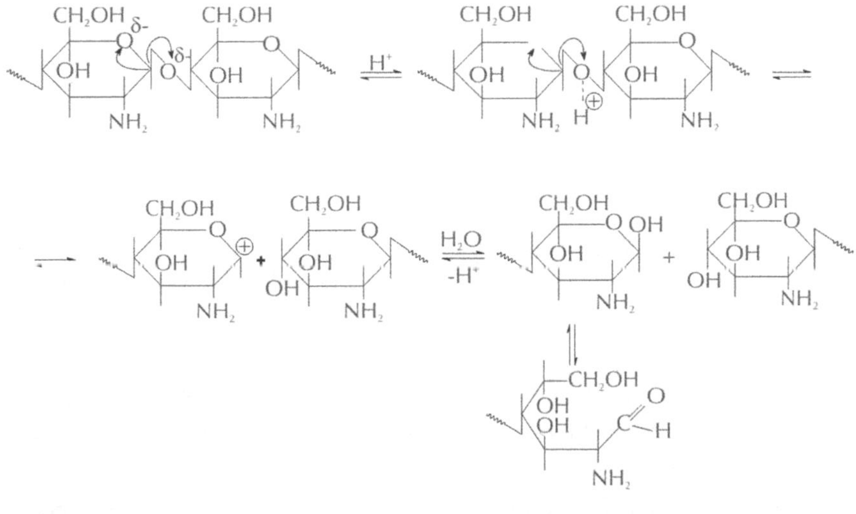

**FIGURE 2**  The scheme of hydrolytic rupture of the glycoside bond.

Direct determination of the reducing carbohydrates concentration during the experiment showed that a prolonged (for 20 days) standing of CHT solutions in acetic acid in the absence of enzymes is not accompanied by any change in their concentration. On the contrary, the standing of CHT solutions in acetic acid in the presence of enzymatic preparations results in a regular increase in reducing carbohydrates concentrations (Fig. 3).

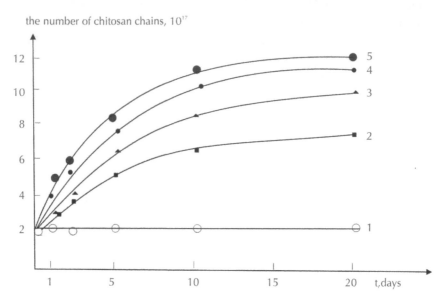

**FIGURE 3**   The dependence of the number of chitosan chains placed in 0.1 ml of 0.2% chitosan solution in acetic acid (1) in the presence of enzymes trypsin (2), lidase (3), collagenase (4) and pepsin (5) on the time of solution standing.

As this takes place, the type of corresponding kinetic dependences doesn't coincide with the dependences given in Fig. 1 both in the course of time and in enzymes efficiency.

In the case of CHT specimens having no contact with enzymes the decrease in intrinsic viscosity observed after its reprecipitation from acetic acid solutions is seen not to result from the hydrolysis process, which is indicated by the time constancy of the reducing carbohydrates concentration. This is most likely to be connected with the destruction of the initial supermolecular structure of the polar CHT disposed to forming intermolecular hydrogen bonds. Correspondingly, the viscosimetry applied to CHT gives only a qualitative idea how destructive transformations take place. In this connection, the direct method of determining the reducing carbohydrates concentration seems to be more correct at investigation of the chitosan enzymatic destruction.

## 11.4 CONCLUSION

Thus in the course of the investigation it has been established that:

(1) in the absence of enzymatic preparations no accumulation of the reducing carbohydrates amount takes place. The observed changes in the values of intrinsic viscosity of CHT which underwent the stage of dissolving in acetic acid are evidently due to rearrangement of the supermolecular polymer structure;

(2) in the presence of enzymatic preparation there occurs the accumulation of the reducing carbohydrates amount caused by the CHT main chain hydrolysis taking place under the action of enzymes.

The work has been carried out due to the financial support of the FAP "Research and scientific – pedagogical personnel of the innovational Russia" (g/k №02.740.11.0648).

## KEYWORDS

- chitosan
- ferricyanide method
- macromolecules aggregates
- viscosimetry

## REFERENCES

1. Ilyina, A. B.; Varlamov, P. V. In: *Chitin and Chitosan: Obtaining, Properties and Application.* Rd. Scryabin, K. G.; Vikhoreva, G. A.; Varlamov P. V., M.: Nauka, **2002**, *79*, 90.
2. Panteleone, D.; Yalpani M., In: *Carbohydrates and Carbohydrate Polymers, Analysis, Biotechnology, Modification, Antiviral, Biomedical and other application.* M. Yalpani, ed. ATL Press. **1993**, 44.
3. Yalpani M.; Panteleone D., *Carbohydr. Res.* **1994**, *256.* 159.
4. Mullagaliev, I. R.; Aktuganov, G. E.; Melentyev I A, *Proceedings of the 7th International Conference "Modern prospects in the investigation of chitin and chitosan."* M.: VNIRO, **2006**, 1079–1084.

5. Vikhoreva, G. A.; Rogovina, S. Z.; Pchelenko, O. M.; Golbraikh S L, Vysocomolec B S, Soed. **2001,** *43(6),* 1079–1084.
6. Fedoseeva, E. N.; Semchicov Yu. D.; Smirnova A L, Vysocomolec. Soed. B. S.; *48(10),* 1930–1935.
7. Severina, S. E., Solovyev, A. G.; *Practical Studies on Biochemistry.* Red. Moscow.: MGU. **1989,** 509.
8. Chernova V.; Kulish, E. I.; Volodina; Kolesov S., *Proceedings of the 9th International Conference "Modern prospects in the investigation of chitin and chitosan."* M.: VNIRO, **2008,** 234.
9. Kulish, E. I.; Chernova V.; Volodina; Torlopov, M. A.; Kolesov S. *Proceedings of the 10th International Conference "Modern prospects in the investigation of chitin and chitosan."* M.: VNIRO, **2010,** 274.
10. Sklyar, A. M.; Gamzazade, A. I.; Rogovina, L. Z.; Tytkova L. V.; Pavlova, S. A.; Rogozhin S. V.; Slonimskiy L. G.; Vysocomolec. A. S. Soed., **1981,** *23(6),* 1396–1403.

# INDEX

# W

Witten-sander clusters, 41

# X

X-ray diffraction (XRD), 127, 129
  patterns, 135

XRD. *See* X-ray diffraction (XRD)

# Z

Zeolites, 241
"Zigzag" carbon nanotubes, 109–110
  electronic structure, 109–110

Milton Keynes UK
Ingram Content Group UK Ltd.
UKHW021355161024
449569UK00055B/1745

9 781774 632826